D0281775

HEINEMANN MODULAR MATHEMATICS *for* EDEXCEL AS AND A-LEVEL

Further Pure Mathematics 1

Geoff Mannall   Michael Kenwood

1 2 3 4 5 6 7

Endorsed by edexcel

heinemann.co.uk
✓ Free online support
✓ Useful weblinks
✓ 24 hour online ordering

01865 888058

Heinemann Educational Publishers
Halley Court, Jordan Hill, Oxford OX2 8EJ
Part of Harcourt Education

Heinemann is the registered trademark of
Harcourt Education Limited

First published 2004

05
10 9 8 7 6 5 4

10-digit ISBN: 0 435 51109 2
13-digit ISBN: 978 0 435511 09 8

Cover design by Gecko Limited

Original design by Geoffrey Wadsley; additional design work by Jim Turner

Typeset and illustrated by Tech-Set Limited, Gateshead, Tyne and Wear

Printed and bound in China through Phoenix Offset

Acknowledgements:

The publisher's and authors' thanks are due to Edexcel for permission to reproduce questions from past examination papers. These are marked with an [E].
The answers have been provided by the authors and are not the responsibility of the examining board.

About this book

This book is designed to provide you with the best preparation possible for your Edexcel FP1 exam. The series authors are senior examiners and exam moderators themselves and have a good understanding of the Edexcel's requirements.

Finding your way around

Finding your way around
To help to find your way around when you are studying and revising use the:

- edge marks (shown on the front page) – these help you to get to the right chapter quickly;
- contents list – this lists the headings that identify key syllabus ideas covered in the book so you can turn straight to them;
- index – if you need to find a topic the bold number shows where to find the main entry on a topic.

Remembering key ideas

We have provided clear explanations of the key ideas and techniques you need throughout the book. Key ideas you need to remember are listed in a summary of key points at the end of each chapter and marked like this in the chapters:

■ $$\sqrt{(-1)} = \mathrm{i}$$

Exercises and exam questions

In this book questions are carefully graded so they increase in difficulty and gradually bring you up to exam standard.

- past exam questions are marked with an [E];
- review exercises on pages 67 and 129 help you practise answering questions from several areas of mathematics at once, as in the real exam;
- exam style practice paper – this is designed to help you prepare for the exam itself;
- answers are included at the end of the book – use them to check your work.

Contents

Inequalities

1

1.1 Further inequalities solved algebraically

Book C1 shows you how to solve inequalities such as:

$$4(x - 3) < 3(3x + 2)$$

and: $$6x^2 + x - 2 \geqslant 0$$

The solution depends on finding critical values of x. These are values of x for the inequality $f(x) > 0$, say, where the sign of $f(x)$ changes between values of x on either side of the critical value.

Example 1

$$4(x - 3) < 3(3x + 2)$$

$$\Rightarrow \quad 4x - 12 < 9x + 6$$

$$\Rightarrow \quad -18 < 5x$$

$$\Rightarrow \quad 5x > -18$$

The critical value here is $x = -\frac{18}{5}$ and the solution required is $x > -\frac{18}{5}$.

Example 2

$$6x^2 + x - 2 \geqslant 0$$

Here $$6x^2 + x - 2 \equiv (3x + 2)(2x - 1)$$

and critical values occur at $x = -\frac{2}{3}$ and $x = \frac{1}{2}$. You take the intervals into which the number line is divided by the critical values and consider the signs of $f(x) \equiv 6x^2 + x - 2$:

	$x < -\frac{2}{3}$	$-\frac{2}{3} < x < \frac{1}{2}$	$x > \frac{1}{2}$
Sign of $f(x)$	$+$	$-$	$+$

As you require $f(x) \geqslant 0$, the solution is $x \leqslant -\frac{2}{3}$ or $x \geqslant \frac{1}{2}$.

The following examples illustrate how other inequalities are solved.

Example 3

Find the set of values of x for which

$$\frac{4x-1}{x+2} < 1$$

Since $$\frac{4x-1}{x+2} < 1 \Rightarrow \frac{4x-1}{x+2} - 1 < 0$$

you should consider

$$\begin{aligned} f(x) &\equiv \frac{4x-1}{x+2} - 1 \\ &= \frac{4x-1-x-2}{x+2} \\ &= \frac{3x-3}{x+2} \end{aligned}$$

A sketch of the graph of $y = \dfrac{3x-3}{x+2}$ looks like this:

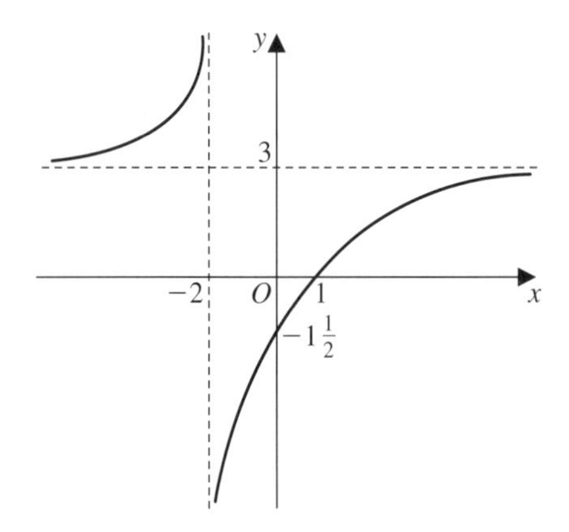

From this you can see that the function changes from positive to negative where the graph crosses the x-axis and also at the vertical asymptote. So, unlike the inequalities in Book C1 where the critical values occurred only where $f(x) = 0$, for some inequalities, the critical values occur where $f(x) = 0$ *and also* at the asymptotes of $y = f(x)$ which are parallel to the y-axis. So the critical values of $f(x)$, where $f(x) \equiv \dfrac{3x-3}{x+2}$, are $x = 1$ and $x = -2$.

	$x < -2$	$-2 < x < 1$	$x > 1$
Sign of $f(x)$	+	−	+

The solution of $\dfrac{4x-1}{x+2} < 1$ is

$$-2 < x < 1$$

Note: A common mistake made by some students is to multiply the inequality by $x+2$, but *crucially* they forget that $x+2$ may be either negative or positive: both possibilities should be considered, particularly since, if negative, the inequality sign is reversed. The approach shown in Example 3 avoids this possible trouble and that is why it is used and recommended.

Example 4

Find the set of values of x for which

$$x^3 \geqslant 7x - 6$$

Since $x^3 \geqslant 7x - 6 \Rightarrow x^3 - 7x + 6 \geqslant 0$

you consider $\quad \mathrm{f}(x) \equiv x^3 - 7x + 6$

By inspection: $\quad \mathrm{f}(1) = 1 - 7 + 6 = 0$

Hence $(x-1)$ is a factor of $\mathrm{f}(x)$.

$$\begin{aligned}\mathrm{f}(x) &\equiv (x-1)(x^2+x-6)\\ &\equiv (x-1)(x-2)(x+3)\end{aligned}$$

The critical values occur at $x = -3$, $x = 1$, $x = 2$.

	$x < -3$	$-3 < x < 1$	$1 < x < 2$	$x > 2$
Sign of $\mathrm{f}(x)$	$-$	$+$	$-$	$+$

The solution set of $x^3 \geqslant 7x - 6$ is

$$-3 \leqslant x \leqslant 1 \text{ or } x \geqslant 2$$

Exercise 1A

In questions 1–15, find the set of values of x for which:

1 $2x - 1 < 4(x-3)$

2 $5(x-2) \geqslant 2(2x+7)$

3 $(x-1)(x-5) > 0$

4 $(x+3)(2x+7) \leqslant 0$

5 $x^2 < 3x + 4$

6 $\dfrac{3}{x-1} > 1$

7 $x^2 < \dfrac{1}{x}$

8 $\dfrac{x}{x-1} < 2$

9 $\dfrac{x-1}{x+1} > 2$

10 $\dfrac{1}{3x^2 - x - 2} < 0$

11 $x^3 - x^2 \geqslant 6x$

12 $\dfrac{x^2 + 10}{x} > 7$

13 $\dfrac{4x - 1}{x + 1} < 2x$

14 $\dfrac{x + 3}{x - 1} < \dfrac{x - 3}{x + 1}$

15 $\dfrac{4x + 8}{x - 1} > 3$

16 Prove that $x + 1$ is a factor of $x^3 - 7x - 6$. Find the values of x for which $x^3 - 7x - 6 \geqslant 0$.

17 Find the complete solution set of

$$\frac{2}{x - 2} > \frac{3}{x + 1}$$

18 Prove that for $x \in \mathbb{R}$, $-1 \leqslant \dfrac{2x}{x^2 + 1} \leqslant 1$.

19 Find the values of x for which

$$\frac{x^2 + 7x + 10}{x + 1} > 2x + 7$$

20 Find the set of values of x for which

$$-1 < \frac{2 - x}{2 + x} \leqslant 1$$

1.2 Solution sets of inequalities from graphs

Book C1 introduces the idea of curve sketching. You may have a calculator with graph sketching facilities; if so, you will find it useful here. If not, just sketch the curves as shown in Book C1.

Example 5
Find the set of values of x for which

$$6 + 5x - 2x^2 - x^3 > 0$$

You can verify that the curve with equation $y = 6 + 5x - 2x^2 - x^3$ cuts the x-axis at the points $(-3, 0)$, $(-1, 0)$ and $(2, 0)$ and the y-axis at the point $(0, 6)$. A sketch of the curve is:

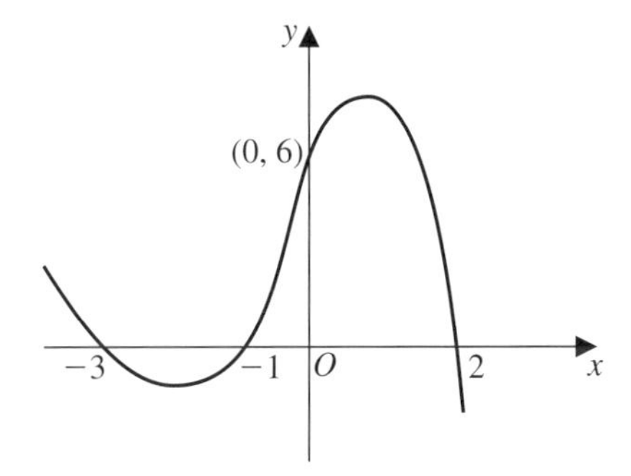

and you can see at once that the curve is above the x-axis for $x < -3$ and for $-1 < x < 2$. The solution set of the inequality is therefore

$$x < -3 \text{ or } -1 < x < 2$$

Note: With a graphical calculator, you type in the equation of the curve, *zoom* in on the intersection points of the curve with the x-axis and read off their coordinates. Record the solution set as shown above *but also produce a rough sketch graph to illustrate your method.*

Example 6

Find the set of values of x for which

$$\frac{2x}{x-1} > x, \quad x \in \mathbb{R}, \quad x \neq 1$$

First sketch the graphs of $y = \dfrac{2x}{x-1}$ and $y = x$ or, using your graphical calculator, type in the two curves so that they are displayed on the same screen. You should obtain a sketch like this:

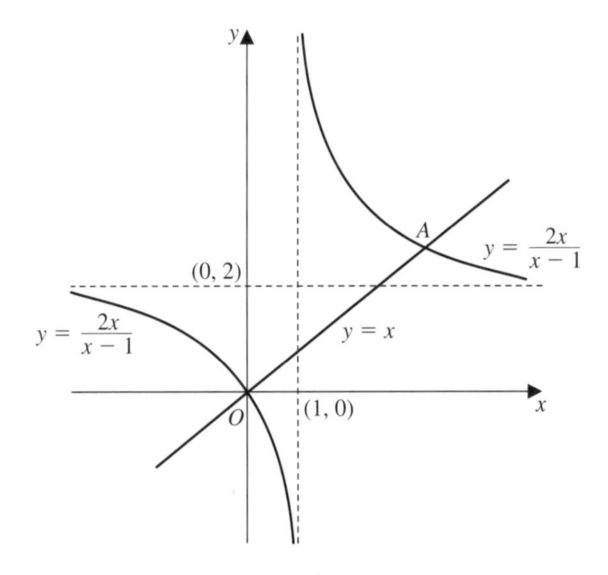

The asymptotes of the curve with equation $y = \frac{2x}{x-1}$ are the lines $x = 1$ and $y = 2$. The curve meets the line with equation $y = x$ at the points O and A, as shown. Zooming in on A, checking by direct calculation, you can show that its coordinates are (3, 3).

The solution set of $\frac{2x}{x-1} > x$ is found by identifying those regions where the curve is 'higher' than the line. You can now see these at once. The solution set is:

$$x < 0 \text{ or } 1 < x < 3$$

Exercise 1B

Many of the questions in Exercise 1A are suitable for solution by a graphical method. You can always check an algebraic solution by sketching a graph too.

Find, by a graphical method, the solution sets of the following inequalities:

1 $\frac{2x}{x+3} < 1$

2 $x(x+1) > x+4$

3 $\frac{5}{x} > x - 4$

4 $\frac{4x}{x+2} > 1$

5 $\frac{1}{x^2 - 6x + 7} > \frac{1}{2}$

6 $4x^3 < 3 + x - 12x^2$

7 $\frac{2x}{x+1} > x$

8 $\frac{x}{2x-1} < 4x$

9 $\frac{1+x}{1-x} < \frac{2-x}{2+x}$

10 $\frac{x^2 + 7x + 10}{x+1} > 2x + 7$

1.3 Inequalities involving the modulus sign

You will already have met modulus functions in Book C3. You should use a graphical approach when solving inequalities that include modulus functions. Avoid techniques such as squaring, because they are unnecessary. The following examples are typical of those set at Advanced level.

Example 7

Find the set of values of x for which

$$|3x+2| < 4x$$

You will recall that the graph of $y = |3x+2|$ is 'V-shaped' and meets the x-axis at the point $(-\frac{2}{3}, 0)$, as shown. The graph of the line $y = 4x$ passes through the origin O and meets the graph of $y = |3x+2|$ where $4x = 3x + 2$ only; that is, at the point $A\,(2, 8)$.

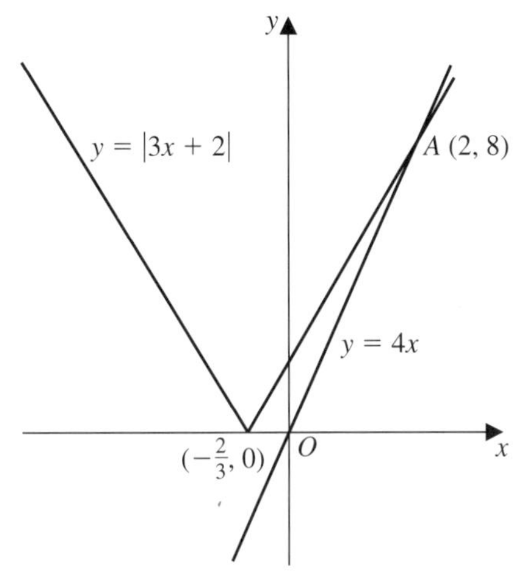

You require the set of values for which $4x$ is greater than $|3x+2|$. That is, $x > 2$ is the solution set.

Example 8

Sketch the graphs of (a) $y = \dfrac{x-1}{x+2}$ (b) $y = \left|\dfrac{x-1}{x+2}\right|$ on separate diagrams.

Using your graph of $y = \left|\dfrac{x-1}{x+2}\right|$, find the set of values of x for which $\left|\dfrac{x-1}{x+2}\right| < 2$.

This is a sketch of the graph of $y = \dfrac{x-1}{x+2}$:

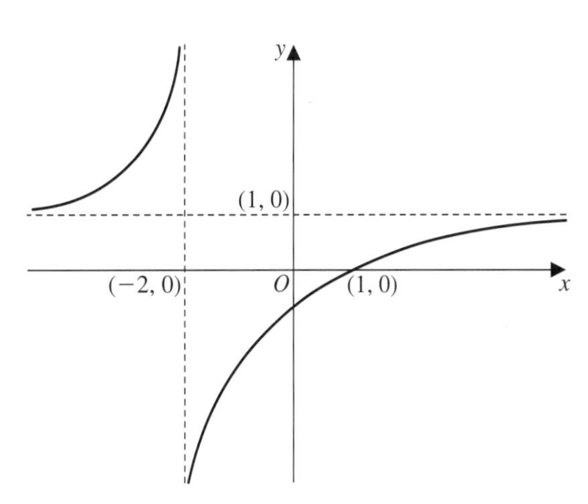

The part of the curve with equation $y = \dfrac{x-1}{x+2}$ for which $y < 0$ is reflected in the x-axis to obtain the curve $y = \left|\dfrac{x-1}{x+2}\right|$:

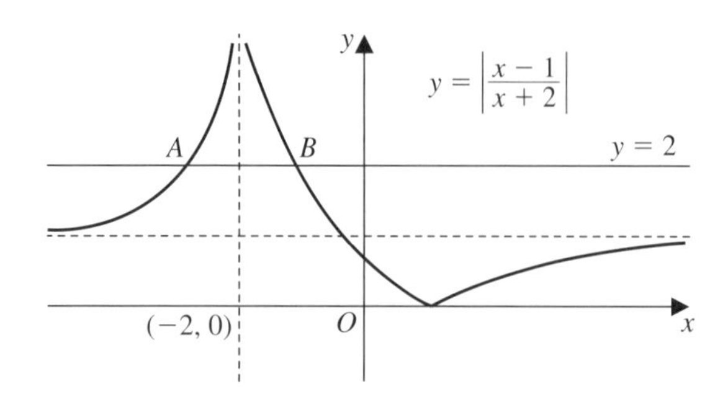

In order to solve this inequality, the line $y = 2$ and the graph of $y = \left|\dfrac{x-1}{x+2}\right|$ are sketched. These two meet at the points A and B, as shown.

For A, $\dfrac{x-1}{x+2} = 2 \Rightarrow x = -5$ and A is $(-5, 2)$.

For B, $\dfrac{-x+1}{x+2} = 2 \Rightarrow x = -1$ and B is $(-1, 2)$.

The solution is the set of values of x for which the graph of $y = \left|\dfrac{x-1}{x+2}\right|$ is 'below' the line $y = 2$.

That is: $x < -5$ or $x > -1$ is the solution set.

Exercise 1C

Find the solution set of:

1 $|x-2| > 4$

2 $|2x+3| < 7$

3 $|3x-4| > 7x$

4 $3|x+1| \geqslant 1-x$

5 $|4-x^2| \leqslant 3$

6 $|x+1| + |x-2| \leqslant 5$

7 $\left|\dfrac{2x}{x-2}\right| < 1$

8 $|x^2-2x| < x$

9 $|8-2x-x^2| < 8$

10 $2x + |x| < 6$

11 Prove that for all real x,

$$\left|\frac{x+1}{x^2+2x+2}\right| \leqslant \frac{1}{2}$$

$\left(\text{Hint: Consider the graph of the curve with equation } y = \frac{x+1}{x^2+2x+2}\right)$

12 The function f is defined for all real x by

$$f(x) = \begin{cases} \frac{1}{x} & \text{for} \quad x > 0 \\ |x| & \text{for} \quad x \leqslant 0 \end{cases}$$

(a) Sketch the graph of f.

(b) Find the set of values of x for which

$$f(x) \leqslant 4$$

SUMMARY OF KEY POINTS

1. Critical values of x for an inequality such as $f(x) > 0$ are those values of x where the sign of $f(x)$ changes for values of x on either side of the critical value.
2. Avoid multiplying an inequality by an expression which could be positive or negative.
3. When using a graphical calculator to solve an inequality, reproduce a rough sketch in your solution to illustrate your method.
4. Use a graphical approach to solve inequalities containing the modulus sign.

Series

2

2.1 Summation of finite series using the method of differences

There is no general method for summing a series. Book C1 shows you how to sum arithmetic series and C2 shows you how to sum geometric series to n terms.

The series $$\sum_{r=1}^{n} 1 = 1 + 1 + 1 + \cdots + 1 = n$$

and $$\sum_{r=1}^{n} r = 1 + 2 + 3 + \cdots + n$$

which is an arithmetic series with first term 1 and common difference 1, were dealt with in Book C1.

Notice that $$\sum_{r=1}^{n} r = 1 + 2 + 3 + \cdots + n \text{ (frontwards)}$$

and also: $$\sum_{r=1}^{n} r = n + (n-1) + (n-2) + \cdots + 1 \text{ (backwards)}$$

Adding: $$2\sum_{r=1}^{n} r = (n+1) + (n+1) + (n+1) + \cdots + (n+1)$$

$$= n(n+1)$$

■ $$\boldsymbol{\sum_{r=1}^{n} r = \tfrac{1}{2}n(n+1)}$$

Here is another way in which you could sum the series

$$1 + 2 + 3 + \cdots$$

to n terms. Consider the identity

$$2r \equiv r(r+1) - (r-1)r$$

If you take successive values $1, 2, 3, \ldots, n$ for r you get:

$$\begin{aligned} 2(1) &= (1)(2) - (0)(1) \\ 2(2) &= (2)(3) - (1)(2) \\ 2(3) &= (3)(4) - (2)(3) \\ &\vdots \\ 2(n-1) &= (n-1)(n) - (n-2)(n-1) \\ 2n &= n(n+1) - (n-1)(n) \end{aligned}$$

Adding up each side gives:

$$2[1 + 2 + 3 + \cdots + (n-1) + n] = n(n+1)$$

All other terms cancel out on the right-hand side. This gives:

$$2\sum_{r=1}^{n} r = n(n+1)$$

That is: $\displaystyle\sum_{r=1}^{n} r = \tfrac{1}{2}n(n+1)$ as before:

This method is called **summing a series by the method of differences**. It is a very elegant and effective way of summing some finite series. Generally, if it is possible to find a function, $f(r)$, such that the rth term u_r of a series can be expressed as

$$u_r \equiv f(r+1) - f(r)$$

then it is easy to find $\displaystyle\sum_{r=1}^{n} u_r$. You have for $r = 1, 2, 3, \ldots, n$:

$$\begin{aligned} u_1 &= f(2) - f(1) \\ u_2 &= f(3) - f(2) \\ u_3 &= f(4) - f(3) \\ &\vdots \\ u_n &= f(n+1) - f(n) \end{aligned}$$

Adding:

■ $$\boldsymbol{\sum_{r=1}^{n} u_r = f(n+1) - f(1)}$$

because all the other terms on the right-hand side cancel out. This is the method of summation of a series using differences. In an examination you will usually be given a hint about what function to consider. Here are some examples.

Example 1

Find $\sum_{r=1}^{n} r^2$.

Consider the identity

$$24r^2 + 2 \equiv (2r+1)^3 - (2r-1)^3$$

and take $r = 1, 2, 3, \ldots, n$.

$$24(1^2) + 2 = 3^3 - 1^3$$
$$24(2^2) + 2 = 5^3 - 3^3$$
$$24(3^2) + 2 = 7^3 - 5^3$$
$$\vdots \quad \vdots \quad \vdots$$
$$24[(n-1)^2] + 2 = (2n-1)^3 - (2n-3)^3$$
$$24[(n)^2] + 2 = (2n+1)^3 - (2n-1)^3$$

Adding you get:

$$24[1^2 + 2^2 + 3^2 + \cdots + (n-1)^2 + n^2] + 2n = (2n+1)^3 - 1^3$$
$$= 8n^3 + 12n^2 + 6n$$

That is: $$24\sum_{r=1}^{n} r^2 + 2n = 8n^3 + 12n^2 + 6n$$

$$24\sum_{r=1}^{n} r^2 = 8n^3 + 12n^2 + 4n = 4n(2n^2 + 3n + 1)$$

$$\sum_{r=1}^{n} r^2 = \frac{n}{6}[2n^2 + 3n + 1]$$

$$= \frac{n}{6}(2n+1)(n+1)$$

■ $$\boldsymbol{\sum_{r=1}^{n} r^2 = \tfrac{1}{6}n(n+1)(2n+1)}$$

Example 2

Find $\sum_{r=1}^{n} r^3$.

Consider the identity

$$4r^3 \equiv r^2(r+1)^2 - (r-1)^2 r^2$$

and take $r = 1, 2, 3, \ldots, n$.

$$\begin{aligned}
4(1^3) &= 1^2(2^2) - 0^2(1^2) \\
4(2^3) &= 2^2(3^2) - 1^2(2^2) \\
4(3^3) &= 3^2(4^2) - 2^2(3^2) \\
&\vdots \\
4(n^3) &= n^2(n+1)^2 - (n-1)^2(n^2)
\end{aligned}$$

Adding you get:

$$4(1^3 + 2^3 + 3^3 + \cdots + n^3) = n^2(n+1)^2$$

That is:

■ $$\boldsymbol{\sum_{r=1}^{n} r^3 = \tfrac{1}{4} n^2(n+1)^2}$$

Notice that since $\sum_{r=1}^{n} r = \frac{1}{2}n(n+1)$, then

$$\sum_{r=1}^{n} r^3 = \left[\sum_{r=1}^{n} r\right]^2$$

Example 3

Find $\sum_{r=1}^{n} r(r+1)$.

Consider the identity

$$3r(r+1) \equiv r(r+1)(r+2) - (r-1)(r)(r+1)$$

and take $r = 1, 2, 3, \ldots, n$.

$$\begin{aligned}
3(1)(2) &= (1)(2)(3) - 0(1)(2) \\
3(2)(3) &= (2)(3)(4) - (1)(2)(3) \\
3(3)(4) &= (3)(4)(5) - (2)(3)(4) \\
&\vdots \\
3(n)(n+1) &= (n)(n+1)(n+2) - (n-1)(n)(n+1)
\end{aligned}$$

Adding you get:

$$3[(1)(2) + (2)(3) + (3)(4) + \cdots + n(n+1)] = n(n+1)(n+2)$$

That is:

$$\sum_{r=1}^{n} r(r+1) = \tfrac{1}{3}n(n+1)(n+2)$$

Exercise 2A

In each case, use the identity given to find the sum to n terms of the given series.

	Identity	Series
1	$\dfrac{1}{r(r+1)} \equiv \dfrac{1}{r} - \dfrac{1}{r+1}$	$\displaystyle\sum_{r=1}^{n} \frac{1}{r(r+1)}$
2	$2r + 1 \equiv (r+1)^2 - r^2$	$\displaystyle\sum_{r=1}^{n} (2r+1)$
3	$\dfrac{2}{4r^2 - 1} \equiv \dfrac{1}{2r-1} - \dfrac{1}{2r+1}$	$\displaystyle\sum_{r=1}^{n} \frac{1}{4r^2 - 1}$
4	$r^2(r+1) - (r-1)^2(r) \equiv 3r^2 - r$	$\displaystyle\sum_{r=1}^{n} r(3r-1)$
5	$\dfrac{r}{r+1} - \dfrac{r-1}{r} \equiv \dfrac{1}{r(r+1)}$	$\displaystyle\sum_{r=1}^{n} \frac{1}{r(r+1)}$
6	$4r(r+1)(r+2) \equiv r(r+1)(r+2)(r+3) - (r-1)(r)(r+1)(r+2)$	$\displaystyle\sum_{r=1}^{n} r(r+1)(r+2)$
7	$\dfrac{2}{r(r+1)(r+2)} \equiv \dfrac{1}{r(r+1)} - \dfrac{1}{(r+1)(r+2)}$	$\displaystyle\sum_{r=1}^{n} \frac{1}{r(r+1)(r+2)}$
8	$\dfrac{2r+1}{r^2(r+1)^2} \equiv \dfrac{1}{r^2} - \dfrac{1}{(r+1)^2}$	$\displaystyle\sum_{r=1}^{n} \frac{2r+1}{r^2(r+1)^2}$

9 Use the identity $(r+1)^3 - r^3 \equiv 3r^2 + 3r + 1$ to find

$$\sum_{r=1}^{n} r(r+1).$$

10 Show that $\dfrac{1}{r!} - \dfrac{1}{(r+1)!} \equiv \dfrac{r}{(r+1)!}$. Hence find $\displaystyle\sum_{r=1}^{n} \frac{r}{(r+1)!}$.

2.2 Summation of finite series using standard results

In addition to formulae for arithmetic and geometric series, you now have from section 2.1 the following results:

- $$\sum_{r=1}^{n} r = \tfrac{1}{2}n(n+1)$$
- $$\sum_{r=1}^{n} r^2 = \tfrac{1}{6}n(n+1)(2n+1)$$
- $$\sum_{r=1}^{n} r^3 = \tfrac{1}{4}n^2(n+1)^2$$

The following examples show how you can use these formulae to sum finite series.

Example 4

Find (a) $\sum_{r=7}^{20} r^2$ (b) $\sum_{r=12}^{25} r^3$.

(a)
$$\begin{aligned}\sum_{r=7}^{20} r^2 &= \sum_{r=1}^{20} r^2 - \sum_{r=1}^{6} r^2 \\ &= \frac{20}{6}(21)(41) - \frac{6}{6}(7)(13) \\ &= 2870 - 91 \\ &= 2779\end{aligned}$$

(b)
$$\begin{aligned}\sum_{r=12}^{25} r^3 &= \sum_{r=1}^{25} r^3 - \sum_{r=1}^{11} r^3 \\ &= \frac{25^2}{4}(26^2) - \frac{11^2}{4}(12^2) \\ &= 105\,625 - 4356 \\ &= 101\,269\end{aligned}$$

Example 5

Prove that $\sum_{r=1}^{n} r(r+1) \equiv \tfrac{1}{3}n(n+1)(n+2)$.

$$\sum_{r=1}^{n} r(r+1) = \sum_{r=1}^{n} (r^2 + r)$$

$$= \sum_{r=1}^{n} r^2 + \sum_{r=1}^{n} r$$

$$= \tfrac{1}{6}n(n+1)(2n+1) + \tfrac{1}{2}n(n+1)$$

$$= \tfrac{1}{6}n(n+1)[(2n+1)+3]$$

$$= \tfrac{1}{6}n(n+1)(2n+4)$$

■ **So:** $$\boldsymbol{\sum_{r=1}^{n} r(r+1) = \tfrac{1}{3}n(n+1)(n+2)}$$

as required.

Example 6

Find $\sum_{r=1}^{n} (6r^2 + 2^r)$.

Now $$\sum_{r=1}^{n} (6r^2 + 2^r) = \sum_{r=1}^{n} 6r^2 + \sum_{r=1}^{n} 2^r$$

$$\sum_{r=1}^{n} 6r^2 = 6\sum_{r=1}^{n} r^2 = n(n+1)(2n+1)$$

$\sum_{r=1}^{n} 2^r$ is a geometric series with first term 2 and common ratio 2, having n terms. The sum of a geometric series:

$$a + ar + \cdots + ar^{n-1}$$

is given by:

$$S = \frac{a(r^n - 1)}{r - 1}$$

where a is the first term and r is the common ratio. (There is more about this in Book P1 page 120.)

Hence: $$\sum_{r=1}^{n} 2^r = \frac{2(2^n - 1)}{2 - 1} = 2^{n+1} - 2$$

So: $$\sum_{r=1}^{n} (6r^2 + 2^r) = n(n+1)(2n+1) + 2^{n+1} - 2$$

Exercise 2B

Evaluate:

1 $\sum_{r=1}^{13} r^2$ **2** $\sum_{r=4}^{11} r^3$ **3** $\sum_{r=11}^{24} r(r+1)$

4 $\sum_{r=1}^{19} r(r+4)$ **5** $\sum_{r=1}^{20} \frac{1}{r(r+1)}$ **6** $\sum_{r=3}^{16} (r+2)^3$

7 $\sum_{r=1}^{14} \left(\frac{3}{4}\right)^r$ **8** $\sum_{r=1}^{20} \frac{1}{(r+3)(r+6)}$ **9** $\sum_{r=4}^{16} (2r-1)^3$

10 $\sum_{r=3}^{23} r(r+1)(r+2)$

11 Prove that $\sum_{r=1}^{n} (2r-1)^2 \equiv \frac{1}{3}n(4n^2-1)$.

12 Prove that $\sum_{r=1}^{n} r(2+r) \equiv \frac{1}{6}n(n+1)(2n+7)$.

13 Find $\sum_{r=1}^{20} \frac{1}{4r^2-1}$.

14 Find $\sum_{r=n}^{2n} r^2$.

15 Given that $f(r) \equiv \frac{1}{r(r+1)}$, show that

$$f(r) - f(r+1) \equiv \frac{2}{r(r+1)(r+2)}$$

Hence find $\sum_{r=5}^{25} \frac{1}{r(r+1)(r+2)}$.

16 Prove that $\sum_{r=1}^{n} \frac{1}{(r+1)(r+2)} = \frac{n}{2(n+2)}$.

17 Find the sum of all even numbers between 2 and 200 inclusive, excluding those which are multiples of 3.

18 Find $\sum_{r=1}^{100} 2r^2 - \sum_{r=1}^{200} r^2$.

19 Find the sum of the series

$$1^2 - 2^2 + 3^2 - 4^2 + \cdots - (2n)^2$$

20 Given that $u_r = r(2r+1) + 2^{r+2}$, find $\sum_{r=1}^{n} u_r$.

SUMMARY OF KEY POINTS

1 $\sum_{r=1}^{n} 1 = n$

2 $\sum_{r=1}^{n} r = \frac{1}{2}n(n+1)$

3 If $u_r \equiv \mathrm{f}(r+1) - \mathrm{f}(r)$, then

$$\sum_{r=1}^{n} u_r = \mathrm{f}(n+1) - \mathrm{f}(1)$$

4 $\sum_{r=1}^{n} r^2 = \frac{1}{6}n(n+1)(2n+1)$

5 $\sum_{r=1}^{n} r^3 = \frac{1}{4}n^2(n+1)^2 = \left[\sum_{r=1}^{n} r\right]^2$

6 $\sum_{r=1}^{n} r(r+1) = \frac{1}{3}n(n+1)(n+2)$

Complex numbers

3

Chapter 2 of Book C1 explains how to solve quadratic equations. One method of doing this is by using the quadratic formula. So for the equation

$$ax^2 + bx + c = 0$$

the solutions are:

$$x = \frac{-b \pm \sqrt{(b^2 - 4ac)}}{2a}$$

This is fine so long as $b^2 - 4ac$ is positive or equal to zero. However, you meet a big problem if $b^2 - 4ac$ is negative.

For example, consider the equation

$$x^2 + 2x + 5 = 0$$

Applying the formula gives:

$$x = \frac{-2 \pm \sqrt{(4 - 20)}}{2} = \frac{-2 \pm \sqrt{(-16)}}{2}$$

So you need the square root of -16. This is the problem. The square root of -16 is not 4, since $4^2 = +16$, and it is not -4 since $(-4)^2 = +16$. Thus you cannot solve this equation at the moment because you cannot find the square root of -16.

3.1 Imaginary numbers

In order to overcome this problem, you need to invent a new number. So mathematicians define the square root of -1 as:

- $\sqrt{(-1)} = \mathrm{i}$

With this definition you can now find $\sqrt{(-16)}$, because

$$\begin{aligned}\sqrt{(-16)} &= \sqrt{(16 \times -1)}\\ &= \sqrt{16} \times \sqrt{(-1)}\\ &= 4 \times \mathrm{i} = 4\mathrm{i}\end{aligned}$$

Similarly, $\sqrt{(-81)} = \sqrt{81} \times \sqrt{(-1)} = 9 \times \mathrm{i} = 9\mathrm{i}$

Also: $\sqrt{(-53)} = \sqrt{53} \times \sqrt{(-1)} = 7.280\mathrm{i}$ (4 s.f.)

A number of the form $b\mathrm{i}$ is called an **imaginary number**. The numbers that you have dealt with in mathematics up until now, for example, 7, -4, $\sqrt{17}$, π, e, are called **real numbers** and the complete set of real numbers is usually denoted by the symbol $\mathbb{R}$. So $5 \in \mathbb{R}$, $-\sqrt{17} \in \mathbb{R}$, $\pi \in \mathbb{R}$, and so on.

Going back to the quadratic equation

$$x^2 + 2x + 5 = 0$$

you can now solve it:

$$\begin{aligned} x &= \frac{-2 \pm \sqrt{(-16)}}{2} \\ &= \frac{-2 \pm \sqrt{16}\sqrt{(-1)}}{2} \\ &= \frac{-2 \pm 4\mathrm{i}}{2} \\ &= -1 \pm 2\mathrm{i} \end{aligned}$$

So the roots of $x^2 + 2x + 5 = 0$ are $x = -1 + 2\mathrm{i}$ and $x = -1 - 2\mathrm{i}$.

You could also solve this equation by completing the square:

$$x^2 + 2x + 5 = 0$$

so:

$$\begin{aligned} x^2 + 2x &= -5 \\ (x+1)^2 - 1 &= -5 \\ (x+1)^2 &= -4 \end{aligned}$$

Now since $\mathrm{i} = \sqrt{(-1)}$ then $\mathbf{i^2 = -1}$.

So: $(x+1)^2 = -4 = 4 \times -1 = 4\mathrm{i}^2$

So:

$$\begin{aligned} (x+1) &= \pm\sqrt{(4\mathrm{i}^2)} \\ (x+1) &= \pm 2\mathrm{i} \end{aligned}$$

So $x = -1 \pm 2\mathrm{i}$, as before.

Example 1

Solve the equation $x^2 + 25 = 0$.

$$x^2 + 25 = 0$$
$$\Rightarrow \quad x^2 = -25$$
$$x = \pm\sqrt{(-25)} = \pm 5\text{i}$$

Example 2

Solve the equation $5x^2 - 2x + 2 = 0$.

$$x = \frac{2 \pm \sqrt{(4 - 40)}}{10}$$
$$= \frac{2 \pm \sqrt{(-36)}}{10}$$
$$= \frac{2 \pm 6\text{i}}{10}$$
$$= \tfrac{1}{5} \pm \tfrac{3}{5}\text{i}$$

Example 3

Solve the equation $2x^2 + 3x + 6 = 0$.

$$x = \frac{-3 \pm \sqrt{(9 - 48)}}{4}$$
$$= \frac{-3 \pm \sqrt{(-39)}}{4}$$
$$= \frac{-3 \pm 6.245\text{i}}{4}$$

So: $\quad x = -0.75 \pm 1.56\text{i}$ (2 d.p.)

3.2 Complex numbers and their manipulation

You will notice from the four quadratic equations solved above that the solutions are all of the form $a \pm b\text{i}$, where $a, b \in \mathbb{R}$. The first solution was $-1 \pm 2\text{i}$, the second $0 \pm 5\text{i}$, then $\frac{1}{5} \pm \frac{3}{5}\text{i}$ and finally $-0.75 \pm 1.56\text{i}$.

- **Any number of the form $a + \text{i}b$ where $a, b \in \mathbb{R}$ is called a *complex number*.**

a is called the **real part** of the number and b is called the **imaginary part** of the number. You write

$$\text{Re}(a + \text{i}b) = a \quad \text{and} \quad \text{Im}(a + \text{i}b) = b$$

Of course, if $b = 0$ the number has no imaginary part (for example $7 + 0i = 7$) and so it is a real number. Real numbers, then, are a subset of complex numbers.

Likewise, if $a = 0$ the number has no real part and is called a **pure imaginary number**. So pure imaginary numbers are also a subset of the set of complex numbers. The set of complex numbers is denoted by $\mathbb{C}$.

Adding and subtracting complex numbers

You can add complex numbers together by adding the real parts and then adding the imaginary parts.

You can subtract one complex number from another by subtracting the real parts and then subtracting the imaginary parts.

So: $$(a + ib) + (c + id) = (a + c) + i(b + d)$$

and: $$(a + ib) - (c + id) = (a - c) + i(b - d)$$

Example 4

Simplify: (a) $(2 + 3i) + (7 - 6i)$ (b) $(25 - 7i) - (13 - 4i)$.

(a) $$(2 + 3i) + (7 - 6i) = (2 + 7) + (3 - 6)i = 9 - 3i$$

(b) $$\begin{aligned}(25 - 7i) - (13 - 4i) &= (25 - 13) + (-7 + 4)i \\ &= 12 - 3i\end{aligned}$$

Multiplying one complex number by another

If you wish to multiply two complex numbers together you must apply the rules of algebra. (These were demonstrated in Book P1, chapter 1.)

So: $$\begin{aligned}(a + ib)(c + id) &= ac + iad + ibc + i^2bd \\ &= ac + i(ad + bc) - bd \text{ (using } i^2 = -1) \\ &= (ac - bd) + i(ad + bc)\end{aligned}$$

Example 5

Multiply $(4 - 3i)$ by $(-7 + 5i)$.

$$\begin{aligned}(4 - 3i)(-7 + 5i) &= -28 + 20i + 21i - 15i^2 \\ &= -28 + 20i + 21i + 15 \\ &= -13 + 41i\end{aligned}$$

Dividing one complex number by another

If you have a complex number $a + ib$ then the complex number $a - ib$ is called the **complex conjugate** of the first, and vice versa. So $(-2 + 5i)$ and $(-2 - 5i)$ are complex conjugates. $-2 + 5i$ is the complex conjugate of $-2 - 5i$, and $-2 - 5i$ is the complex conjugate of $-2 + 5i$.

- **If z is a complex number, its complex conjugate is denoted by z^*.**

When you multiply a complex number by its conjugate the result is always a *real* number. For example:

$$\begin{aligned}(a + ib)(a - ib) &= a^2 - abi + abi - i^2b^2 \\ &= a^2 - abi + abi + b^2 \\ &= a^2 + b^2\end{aligned}$$

Complex conjugates are useful when you need to divide one complex number by another.

Example 6

Express $\dfrac{2 + 3i}{1 - 2i}$ in the form $a + bi$ where $a, b \in \mathbb{R}$.

Multiplying both the numerator and the denominator by the complex conjugate of $1 - 2i$ gives:

$$\begin{aligned}\frac{2 + 3i}{1 - 2i} \times \frac{1 + 2i}{1 + 2i} &= \frac{2 + 4i + 3i + 6i^2}{1 + 2i - 2i - 4i^2} \\ &= \frac{2 + 7i - 6}{1 + 4} \\ &= \frac{-4 + 7i}{5} \\ &= -\tfrac{4}{5} + \tfrac{7}{5}i\end{aligned}$$

So the technique with division is to multiply both the numerator and the denominator by the complex conjugate of the denominator.

Example 7

Show that $\frac{1+2i}{3-i}+\frac{1-2i}{3+i}$ is real.

$$\frac{1+2i}{3-i}+\frac{1-2i}{3+i}=\left(\frac{1+2i}{3-i}\times\frac{3+i}{3+i}\right)+\left(\frac{1-2i}{3+i}\times\frac{3-i}{3-i}\right)$$

$$=\left(\frac{3+6i+i+2i^2}{9-i^2}\right)+\left(\frac{3-6i-i+2i^2}{9-i^2}\right)$$

$$=\frac{3+7i+2i^2+3-7i+2i^2}{9-i^2}$$

$$=\frac{6+4i^2}{9-i^2}$$

$$=\frac{6-4}{9+1}$$

$$=\tfrac{2}{10}=\tfrac{1}{5}\text{, which is real.}$$

Exercise 3A

1 Express in terms of i:

(a) $\sqrt{(-64)}$ (b) $\sqrt{(-7)}$
(c) $\sqrt{16}-\sqrt{(-81)}$ (d) $3-\sqrt{(-25)}$
(e) $\sqrt{(-100)}-\sqrt{(-49)}$

2 Simplify:

(a) i^3 (b) i^7
(c) i^{-9} (d) $i(2i-3i^3)$
(e) $(i+2i^2)(3-i)$

3 Write in the form $a+ib$, where $a, b \in \mathbb{R}$:

(a) $2i(5-2i)$ (b) $(2+i)^2$
(c) $(4-i)^5$ (d) $(1+2i)^2+(3-i)^3$
(e) $(1+i)^2-3(2-i)^3$

4 Find z^* given that $z =$

(a) $2+4i$ (b) $3-6i$
(c) $-5+2i$ (d) $-7-3i$
(e) $2i-4$ (f) 6
(g) $3i$ (h) $-3i+7$

5 Simplify:

(a) $(2 + 3\text{i}) + (4 - 7\text{i})$ (b) $(-3 + 5\text{i}) + (-6 - 7\text{i})$

(c) $(-7 - 10\text{i}) + (2 - 3\text{i})$ (d) $(2 + 4\text{i}) - (3 - 6\text{i})$

(e) $(-3 + 5\text{i}) - (-7 + 4\text{i})$ (f) $(-9 - 6\text{i}) - (-8 - 9\text{i})$

(g) $(6 - 3\text{i}) - (8 - 5\text{i})$

6 Express in the form $a + \text{i}b$, where $a, b \in \mathbb{R}$:

(a) $(2 + \text{i})(3 - \text{i})$ (b) $(-3 - 4\text{i})(2 - 7\text{i})$

(c) $(5 + 2\text{i})(-3 + 4\text{i})$ (d) $(1 - 5\text{i})^2$

(e) $(2 - \text{i})^3$ (f) $(1 + \text{i})(2 - \text{i})(\text{i} + 3)$

(g) $\text{i}(3 - 7\text{i})(2 - \text{i})$

7 Express in the form $a + \text{i}b$, where $a, b \in \mathbb{R}$:

(a) $\dfrac{2 - 7\text{i}}{1 + 2\text{i}}$ (b) $\dfrac{1 + 2\text{i}}{3 - \text{i}}$

(c) $\dfrac{1 + 2\text{i}}{3 + 4\text{i}}$ (d) $\dfrac{1}{1 + 2\text{i}}$

(e) $\dfrac{2 + 3\text{i}}{2 - 3\text{i}}$ (f) $\dfrac{5 + \text{i}}{\text{i} - 3}$

(g) $\dfrac{6}{4\text{i} - 3}$ (h) $\dfrac{1}{(\text{i} + 2)(1 - 2\text{i})}$

8 Solve:

(a) $x^2 + 25 = 0$ (b) $x^3 + 64x = 0$

(c) $x^2 - 4x + 5 = 0$ (d) $x^2 + 6x + 10 = 0$

(e) $x^2 + 29 = 4x$ (f) $2x^2 + 3x + 7 = 0$

(g) $3x^2 + 2x + 1 = 0$ (h) $3x^2 - 2x + 2 = 0$

9 Express in the form $a + \text{i}b$, where $a, b \in \mathbb{R}$:

(a) $\dfrac{1}{1 + 2\text{i}} + \dfrac{1}{1 - 2\text{i}}$ (b) $\dfrac{1}{2 + \text{i}} - \dfrac{1}{1 + 5\text{i}}$

(c) $5 - 4\text{i} + \dfrac{5}{3 - 4\text{i}}$

10 Given that $z = -1 + 3\text{i}$, express $z + \dfrac{2}{z}$ in the form $a + \text{i}b$, where $a, b \in \mathbb{R}$.

11 Given that $T = \dfrac{x - \mathrm{i}y}{x + \mathrm{i}y}$, where $x, y, T \in \mathbb{R}$, show that

$$\frac{1 + T^2}{2T} = \frac{x^2 - y^2}{x^2 + y^2}$$

12 Show that the complex number $\dfrac{2 + 3\mathrm{i}}{5 + \mathrm{i}}$ can be expressed in the form $\lambda(1 + \mathrm{i})$, where λ is real.
State the value of λ.

Hence, or otherwise, show that $\left(\dfrac{2 + 3\mathrm{i}}{5 + \mathrm{i}}\right)^4$ is real and determine its value. [E]

3.3 The Argand diagram

Every pair of coordinates (x, y) can be plotted as a unique point on a pair of cartesian axes. So $(2, 3)$, $(1, 4)$, $(3, 2)$, $(-5, 7)$ are all distinct points when plotted.

Now the complex numbers $2 + 3\mathrm{i}$, $1 + 4\mathrm{i}$, $3 + 2\mathrm{i}$, $-5 + 7\mathrm{i}$ are all distinct. So if you draw a pair of cartesian axes and take the x-axis as the real axis and the y-axis as the imaginary axis, then for each complex number $z = x + \mathrm{i}y$ there is a unique point (x, y) that you can plot using the axes. So the complex number $z_1 = 5 + 3\mathrm{i}$ can be represented as the point $(5, 3)$, the number $z_2 = 4 - 2\mathrm{i}$ can be represented as $(4, -2)$, and so on.

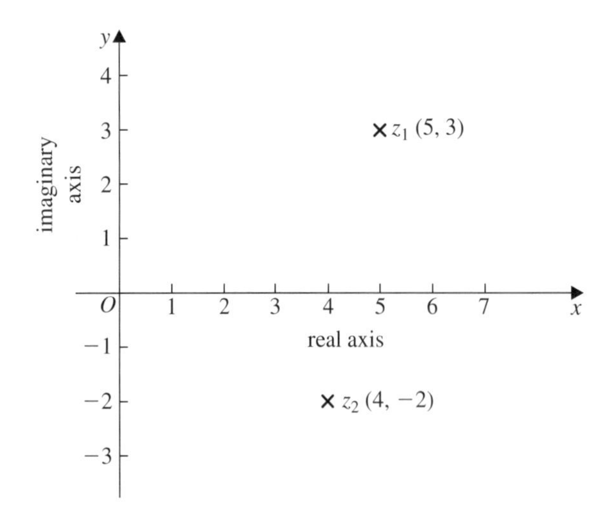

Each complex number besides being represented as a point can also be represented as a vector:

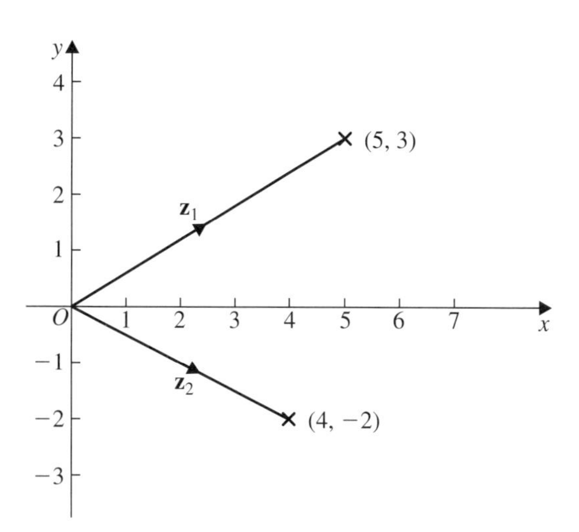

Of course, in this vector method of representation, any vector of the same magnitude as $\mathbf{z}_1$ and in the same direction as $\mathbf{z}_1$ could represent z_1, and the same with z_2, etc. The diagram used to represent complex numbers like this is called an **Argand diagram**.

If the vector method is used to represent complex numbers, then the parallelogram or triangle laws for the addition and subtraction of vectors can be used to find the representation of $z_1 + z_2$ and $z_1 - z_2$ on the Argand diagram:

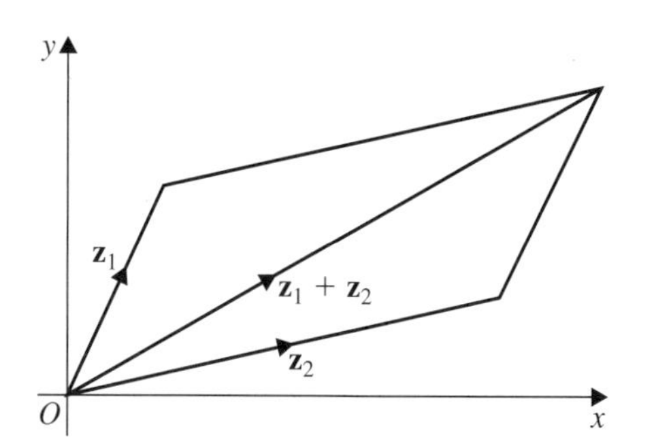

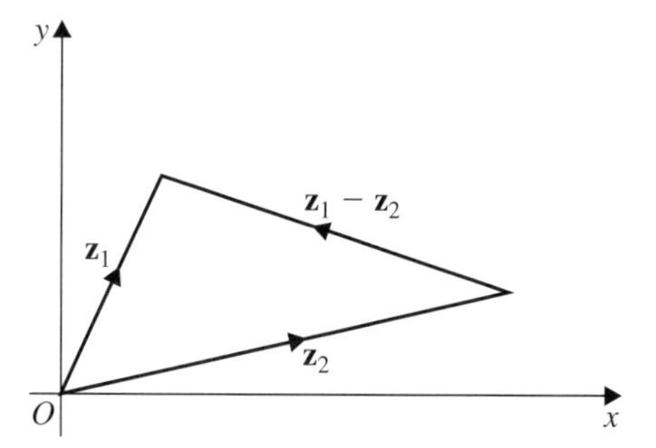

3.4 Modulus and argument of a complex number

The **modulus** of a complex number is defined as the length of the corresponding vector on the Argand diagram. So:

- **the modulus of $z = x + \mathrm{i}y$ is**

$$|z| = \sqrt{(x^2 + y^2)}$$

The **argument** of a complex number is the angle which the vector representing that number makes with the positive x-axis.

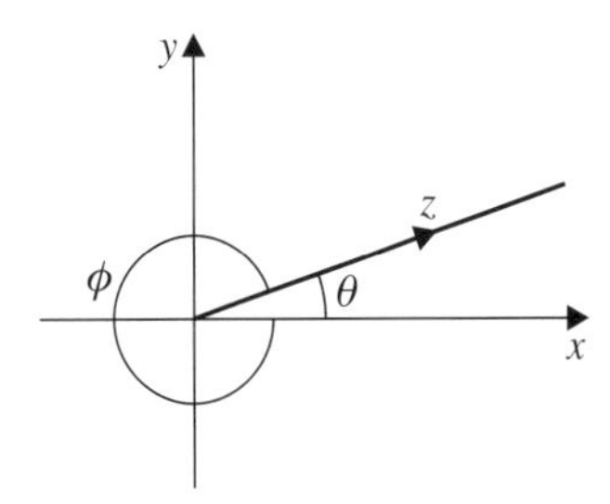

This definition needs a bit of thought though, since as you can see from the diagram the angle could be either θ or ϕ. Moreover, if θ is measured in degrees then $\theta^\circ + 360^\circ$, $\theta^\circ + 720^\circ$, $\theta^\circ + 1080^\circ$, etc. are each an angle which the vector representing z makes with the positive x-axis. Likewise, if θ is measured in radians, $\theta + 2\pi$, $\theta + 4\pi$, $\theta + 6\pi$, etc. could be the angle.

To get over this problem we define the **principal argument** of z to be the angle α which the vector makes with the positive x-axis and such that $-\pi < \alpha \leqslant \pi$ (if α is in radians) or $-180^\circ < \alpha \leqslant 180^\circ$ (if α is in degrees).

It is essential when you are asked to find the argument of a complex number that you draw the relevant vector on an Argand diagram. In this way you will know in which quadrant the vector lies and so you should be able to obtain the correct principal argument. You should be very careful, because a calculator will frequently not give you the *principal* argument.

In this book, and in the examination, you will often be asked for the argument of a complex number. In all cases it is the principal argument that you are being asked to find.

- **If z is a complex number, then the principal argument is often written as arg z.**

Example 8

Find the modulus and argument of:

(a) $1 + \mathrm{i}$ (b) $3 - 4\mathrm{i}$

(c) $-2 - 3\mathrm{i}$ (d) $-5 + 4\mathrm{i}$.

(a)

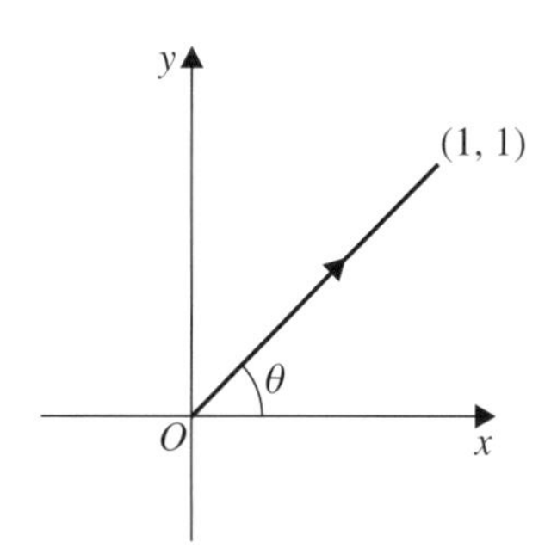

A vector representing $1 + \mathrm{i}$ goes from the origin to $(1, 1)$.

$$|1 + \mathrm{i}| = \sqrt{(1^2 + 1^2)} = \sqrt{2}$$

$\arg(1 + \mathrm{i})$ is the angle θ marked on the diagram.

Now by trigonometry: $\tan\theta = \frac{1}{1}$ and $\theta = \frac{\pi}{4}$

So: $|z| = |1 + \mathrm{i}| = \sqrt{2}, \arg z = \frac{\pi}{4}$

(b)

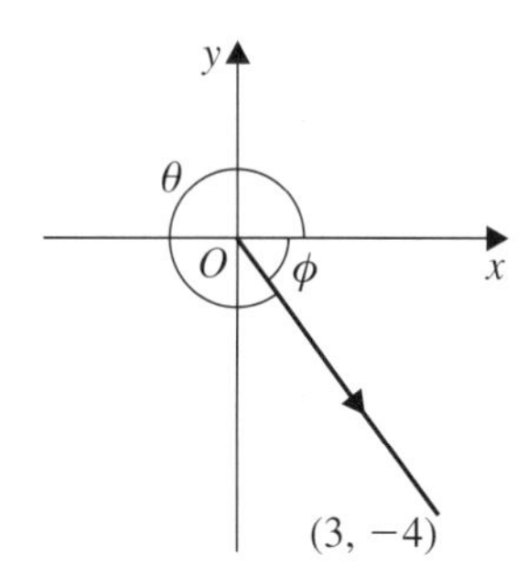

$$|3 - 4\mathrm{i}| = \sqrt{[3^2 + (-4)^2]} = \sqrt{25} = 5$$

In this case ϕ is the principal argument, because it is the angle made with the positive x-axis and it is acute. So $\arg(3 - 4\mathrm{i})$ is ϕ as shown on the diagram.

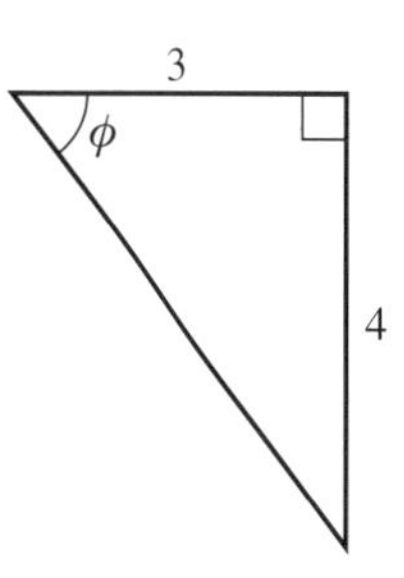

Now in the triangle $\tan\phi = \frac{4}{3}$

So: $\phi = 0.9273$ radians (4 s.f.)

Thus $\arg(3 - 4\mathrm{i}) = -0.9273$ radians (4 s.f.)

(c)

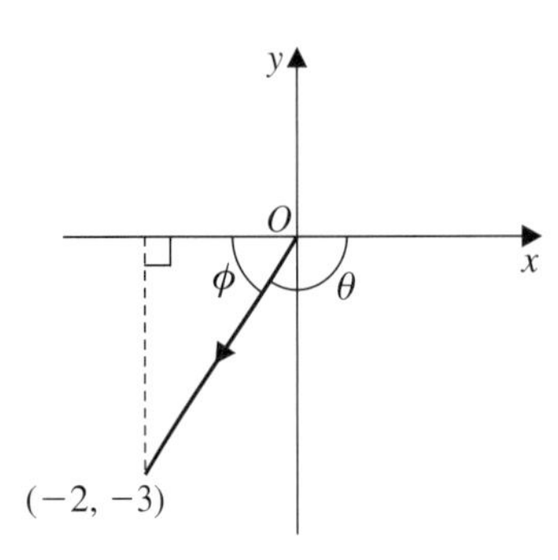

$$|-2-3\mathrm{i}| = \sqrt{[(-2)^2+(-3)^2]} = \sqrt{(4+9)} = \sqrt{13}$$

Since θ on the diagram is the angle required (because $-\pi < \theta \leqslant \pi$) and since this is obtuse, you need to work in the right-angled triangle containing ϕ:

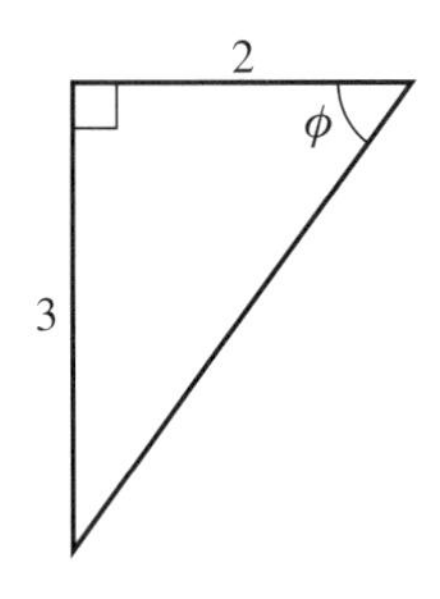

$$\tan\phi = \tfrac{3}{2}$$

$$\Rightarrow \quad \phi = 0.9828 \text{ radians (4 s.f.)}$$

$$\pi - 0.9828 = 2.159 \text{ radians (4 s.f.)}$$

So $\quad \arg(-2-3\mathrm{i}) = -2.159$ radians (4 s.f.)

(d)

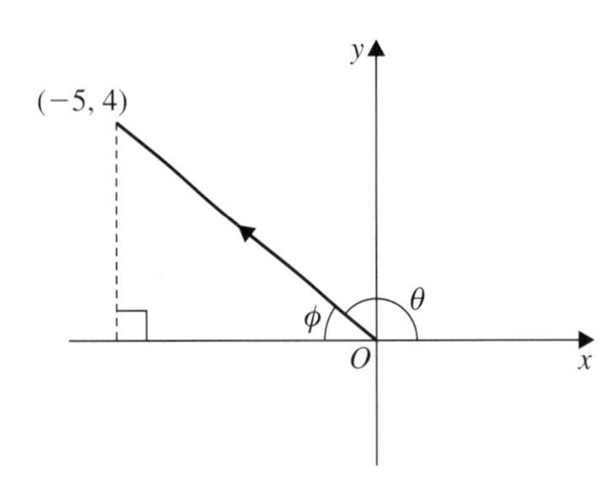

$$|-5+4\mathrm{i}| = \sqrt{[(-5)^2+4^2]} = \sqrt{(25+16)} = \sqrt{41}$$

Since θ on the diagram is the angle required and since θ is obtuse, you must work in the right-angled triangle containing ϕ:

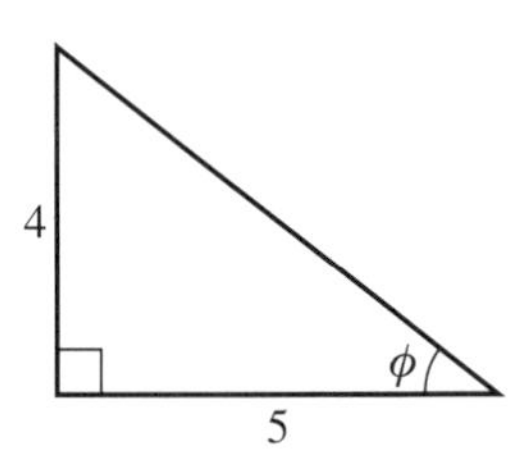

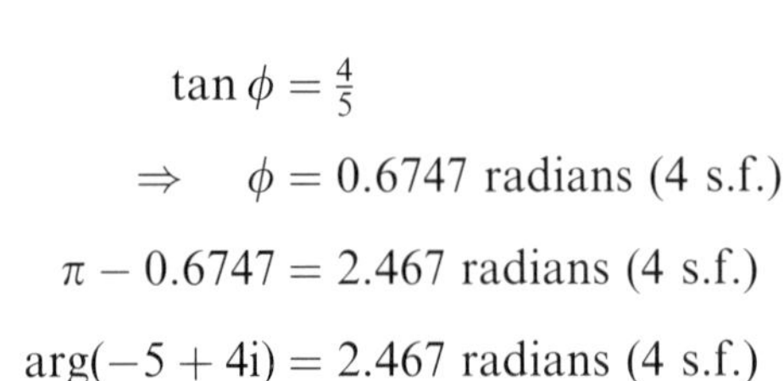

$$\tan\phi = \tfrac{4}{5}$$

$$\Rightarrow \quad \phi = 0.6747 \text{ radians (4 s.f.)}$$

$$\pi - 0.6747 = 2.467 \text{ radians (4 s.f.)}$$

So $\quad \arg(-5+4\mathrm{i}) = 2.467$ radians (4 s.f.)

Example 9

Express $\dfrac{2-\mathrm{i}}{3+\mathrm{i}}$ in the form $a+\mathrm{i}b$ where $a, b \in \mathbb{R}$. Hence find $\left|\dfrac{2-\mathrm{i}}{3+\mathrm{i}}\right|$ and $\arg\left(\dfrac{2-\mathrm{i}}{3+\mathrm{i}}\right)$ and show $2-\mathrm{i}$, $3+\mathrm{i}$ and $\dfrac{2-\mathrm{i}}{3+\mathrm{i}}$ on an Argand diagram.

$$\frac{2-\mathrm{i}}{3+\mathrm{i}} = \frac{2-\mathrm{i}}{3+\mathrm{i}} \times \frac{3-\mathrm{i}}{3-\mathrm{i}}$$

$$= \frac{6-5\mathrm{i}+\mathrm{i}^2}{9+1}$$

$$= \frac{5-5\mathrm{i}}{10}$$

$$= \tfrac{1}{2} - \tfrac{1}{2}\mathrm{i}$$

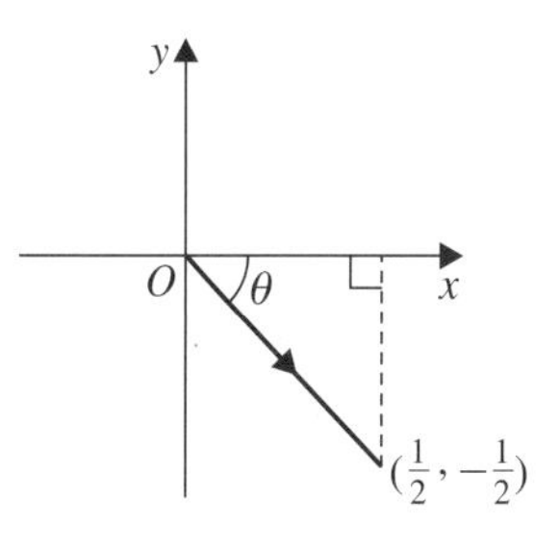

$$\left|\tfrac{1}{2} - \tfrac{1}{2}\mathrm{i}\right| = \sqrt{\left[\left(\tfrac{1}{2}\right)^2 + \left(-\tfrac{1}{2}\right)^2\right]}$$

$$= \sqrt{\left(\tfrac{1}{4} + \tfrac{1}{4}\right)} = \tfrac{1}{\sqrt{2}}$$

$\arg\left(\tfrac{1}{2} - \tfrac{1}{2}\mathrm{i}\right) = \theta$ on the diagram.

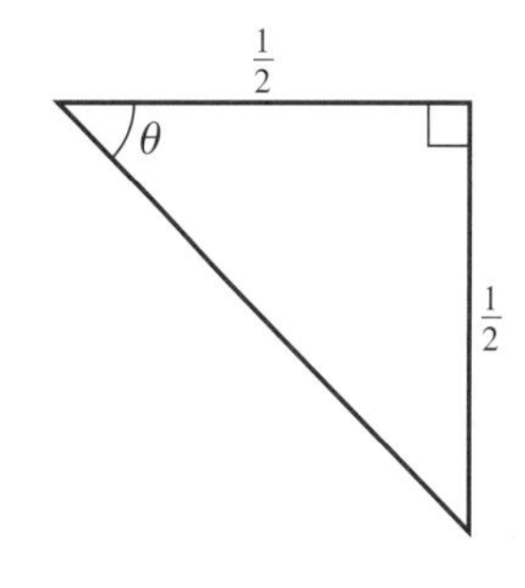

$$\tan\theta = \frac{\frac{1}{2}}{\frac{1}{2}} = 1$$

$$\theta = \frac{\pi}{4}$$

So $\qquad \arg\left(\tfrac{1}{2} - \tfrac{1}{2}\mathrm{i}\right) = -\dfrac{\pi}{4}$

Finally, the Argand diagram looks like this:

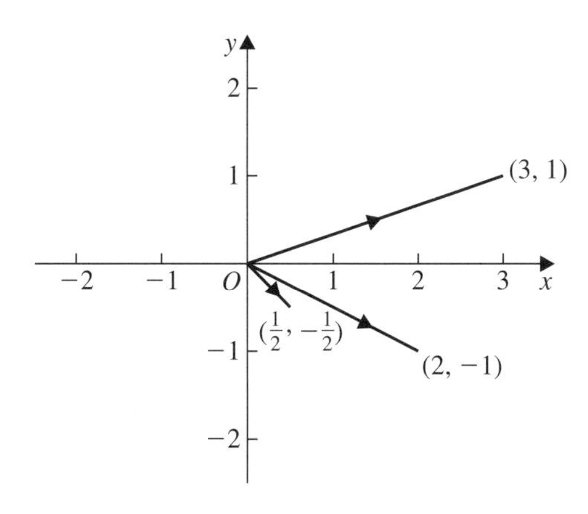

Exercise 3B

1 Represent the following on an Argand diagram:

(a) $2 + 3\text{i}$ (b) $-4 + 5\text{i}$ (c) $6 + 2\text{i}$

(d) $-3 - 2\text{i}$ (e) $2 - 4\text{i}$ (f) $-7 + 3\text{i}$

(g) $2\text{i} - 5$ (h) -5i (i) $-3\text{i} + 4$

(j) $12\text{i} - 5$

2 Find $|z|$ and $\arg z$, in radians to 3 significant figures, where $z =$

(a) $3 - 2\text{i}$ (b) $3 + \text{i}$ (c) 6i

(d) -5 (e) $-2 + \text{i}$ (f) $1 - 3\text{i}$

(g) $\text{i}\sqrt{3} + 1$ (h) $-5 + 12\text{i}$ (i) $\dfrac{5}{1 - \text{i}\sqrt{3}}$

(j) $\dfrac{2}{\sqrt{5} + \text{i}}$ (k) $(2 + \text{i})(3 - 2\text{i})$ (l) $\dfrac{1 + \text{i}}{2 - \text{i}}$

3 $z_1 = 3 + 4\text{i}$, $z_2 = 5 - 12\text{i}$, $z_3 = -1 - \text{i}$, $z_4 = -7 + 24\text{i}$. Find:

(a) $|z_1|$, $|z_2|$, $|z_3|$, $|z_4|$

(b) $\arg z_1$, $\arg z_2$, $\arg z_3$, $\arg z_4$

giving your answers in degrees to 1 decimal place.

4 Given that $a, b \in \mathbb{R}$, express each of the following in the form $a + \text{i}b$:

(a) $\dfrac{3 + \text{i}}{2 - \text{i}}$ (b) $3\text{i}^3 - 6\text{i}^6$ (c) $(1 + \text{i})^4$

Find the modulus and the argument, in radians to 2 decimal places, of each answer.

5 Given that $z = 2 - \text{i}$, find z^2 and z^3.

Find $|z|$, $|z^2|$, $|z^3|$.

Find also, in radians to 2 decimal places, $\arg z$, $\arg z^2$, $\arg z^3$.

6 Find the modulus and the argument, in radians in terms of π, of:

(a) $z_1 = \dfrac{1 + \text{i}}{1 - \text{i}}$ (b) $z_2 = \dfrac{\sqrt{2}}{1 - \text{i}}$ (c) $z_3 = \left(\dfrac{1 + \text{i}}{1 - \text{i}}\right)^2$

Plot z_1, z_2 and $z_1 + z_2$ on a Argand diagram.

Deduce that $\tan \dfrac{3\pi}{8} = 1 + \sqrt{2}$.

7 Express $(3+2i)^2$ and $\dfrac{1}{(3+2i)^2}$ in the form $a+ib$, $a, b \in \mathbb{R}$.

Find $|(3+2i)^2|$ and $\left|\dfrac{1}{(3+2i)^2}\right|$. Find also, in radians to 2 decimal places, $\arg(3+2i)^2$ and $\arg\dfrac{1}{(3+2i)^2}$.

8 Given that $z = \cos\theta + i\sin\theta$ show that

$$\frac{2}{1+z} = 1 - i\tan\tfrac{1}{2}\theta$$

9 Given that $z = -1+3i$, express $z + \dfrac{2}{z}$ in the form $a+ib$ where $a, b \in \mathbb{R}$.

Find $\left|z + \dfrac{2}{z}\right|$.

10 Find (i) the modulus (ii) the argument, in radians to 2 decimal places, of:

(a) $(-2+3i)(1+i\sqrt{3})$

(b) $\dfrac{-2+3i}{1+i\sqrt{3}}$

(c) $[(-2+3i)+(1+i\sqrt{3})]^2$

11 Given that $z_1 = 3+4i$, $z_2 = -1+2i$ place z_1, z_2, z_1+z_2, z_2-z_1 on an Argand diagram. Express $\dfrac{z_2+z_1}{z_2-z_1}$ in the form $p+iq$ where $p, q \in \mathbb{R}$.

12 Given that $z_1 = 3-2i$ and $z_2 = -1+7i$ show that

$$|z_1+z_2| < |z_1-z_2| < |z_1|+|z_2|$$

13 The complex number z has modulus 4 and argument $\dfrac{\pi}{3}$. Find in the form $a+ib$, where $a, b \in \mathbb{R}$:

(a) z^2 (b) $\dfrac{1}{z}$ (c) $i^3 z$

14 In an Argand diagram O is the origin, P represents the number $7-i$ and Q represents the number $12+4i$.

(a) Show that $\triangle OPQ$ is isosceles.

(b) Calculate the size of $\angle OPQ$, giving your answer to the nearest degree.

15
$$\frac{1}{w} = 1 - 2z - z^2$$

Given that $z = -1 + 2\text{i}$,

(a) express w in the form $a + \text{i}b$, where $a, b \in \mathbb{R}$

(b) find $\arg z$ in radians to 2 decimal places.

16
$$z = -10 + 9\text{i}$$

(a) Calculate $\arg z$ to the nearest $0.1°$.

(b) Find the complex number w, given that $wz = -11 + 28\text{i}$.

(c) Calculate $|w|$, giving your answer to one decimal place.

17 Given that $z = -\frac{1}{2} + \frac{\text{i}\sqrt{3}}{2}$, find z^2.

Show that $1 + z + z^2 = 0$.

Place 1, z and z^2 on an Argand diagram and show that the points representing them, when joined, form an equilateral triangle.

18 $\text{f}(x) \equiv x^3 + kx^2 + 9x + 13$, where $k \in \mathbb{R}$.

Given that $\text{f}(-1) = 0$, find the value of k and show that the equation $\text{f}(x) = 0$ has one real and two complex roots. Display the roots on an Argand diagram.

19 Given that $z = \frac{\sqrt{3} + \text{i}}{1 + \text{i}\sqrt{3}}$, find $|z|$ and $\arg z$, giving $\arg z$ in radians to 2 decimal places. Find z^2 in the form $a + \text{i}b$ where $a, b \in \mathbb{R}$. Hence place z and z^2 on the same Argand diagram.

20 Given that $z_1 = 3 - 4\text{i}$ and $z_2 = 12 + 5\text{i}$,

(a) express z_1z_2 and $\frac{z_2}{z_1}$ in the form $a + \text{i}b$, where $a, b \in \mathbb{R}$

(b) place z_1, z_2, z_1z_2 and $\frac{z_2}{z_1}$ on the same Argand diagram.

21 Find the modulus and argument of $z_1 = \sqrt{3} - \text{i}$ and $z_2 = -\sqrt{2} + \text{i}\sqrt{2}$.

Express $\frac{z_2}{z_1}$ in the form $a + \text{i}b$ where $a, b \in \mathbb{R}$, and place z_1, z_2, $z_1 + z_2$ and $\frac{z_2}{z_1}$ on an Argand diagram.

22 Given that $z_1 = -1 + i\sqrt{3}$ and $z_2 = \sqrt{3} + i$, find $\arg z_1$ and $\arg z_2$.

Express $\frac{z_1}{z_2}$ in the form $a + ib$, where a and b are real, and

hence find $\arg \frac{z_1}{z_2}$.

Verify that $\arg \frac{z_1}{z_2} = \arg z_1 - \arg z_2$. [E]

23 Given that $z = 2 - i$, show that $z^2 = 3 - 4i$.
Hence, or otherwise, find the roots, z_1 and z_2, of the equation

$$(z + i)^2 = 3 - 4i$$

Display these roots on an Argand diagram.
(a) Deduce that $|z_1 - z_2| = 2\sqrt{5}$.
(b) Find the value of $\arg(z_1 + z_2)$. [E]

24 Find the modulus and argument of each of the complex numbers z_1 and z_2, where

$$z_1 = 1 + i, z_2 = \sqrt{3} - i$$

Use $\arg \frac{z_1}{z_2} = \arg z_1 - \arg z_2$ to show that

$$\arg \frac{z_1}{z_2} = \frac{5\pi}{12}$$ [E]

25 The complex numbers $z_1 = 2 + 2i$ and $z_2 = 1 + 3i$ are represented on an Argand diagram by the points P and Q respectively.
(a) Display z_1 and z_2 on the same Argand diagram.
(b) Calculate $|z_1|$, $|z_2|$ and the length of PQ.
(c) Hence show that:
(i) $\triangle OPQ$, where O is the origin, is right angled
(ii) $\arg z_2 - \arg z_1 = \arctan \frac{1}{2}$.
(d) Given that $OPQR$ is a rectangle in the Argand diagram, state the complex number z_3 represented by the point R. [E]

26 The complex numbers z and w are given by

$$z = \frac{5 - 10i}{2 - i} \quad \text{and} \quad w = iz$$

(a) Obtain z and w in the form $p + \mathrm{i}q$ where p and q are real numbers.
(b) Show z and w on an Argand diagram.
The origin O and the points representing z and w are the vertices of a triangle.
(c) Show that this triangle is isosceles and state the size of the angle between the equal sides. [E]

3.5 Modulus–argument form of a complex number

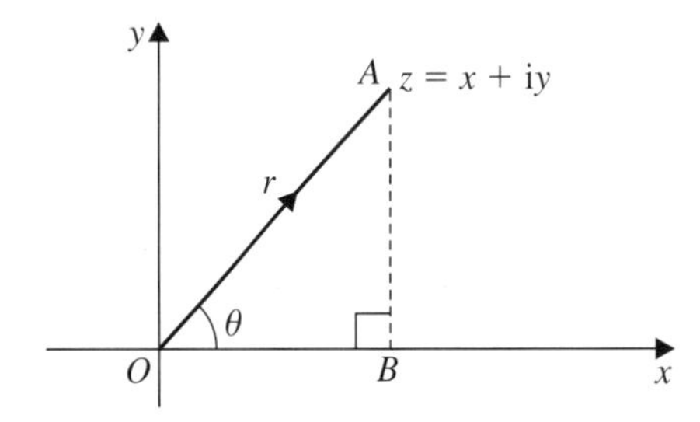

Suppose the complex number $z = x + \mathrm{i}y$ is represented on an Argand diagram by the vector $\overrightarrow{OA}$ and B is the foot of the perpendicular from A to the x-axis.

Then $\quad OB = x \quad \text{and} \quad BA = y$

But $\quad OA = |\overrightarrow{OA}| = |z| = r$

and $\quad \angle BOA = \theta$

So: $\quad \dfrac{OB}{OA} = \cos\theta$

That is: $\quad \dfrac{OB}{r} = \cos\theta \Rightarrow OB = r\cos\theta$

Also $\quad \dfrac{AB}{OA} = \sin\theta$

That is: $\quad \dfrac{AB}{r} = \sin\theta \Rightarrow AB = r\sin\theta$

So the complex number $z = x + \mathrm{i}y$ can be written

$$z = r\cos\theta + \mathrm{i}r\sin\theta$$

or

■ $\boldsymbol{z = r(\cos\theta + \mathrm{i}\sin\theta)}$

This is called the **modulus–argument form** of a complex number z.

Example 10

Express $z = 5 + 12\text{i}$ in the form $r(\cos\theta + \text{i}\sin\theta)$ where $r = |z|$ and $\theta = \arg z$.

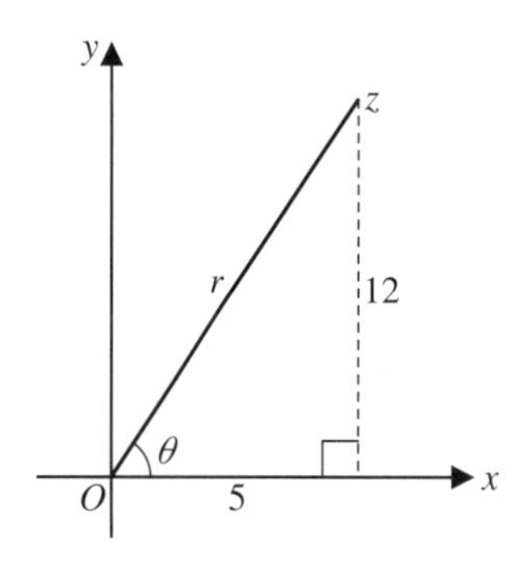

$$z = 5 + 12\text{i}$$

$$r = |z| = \sqrt{(5^2 + 12^2)} = \sqrt{(25 + 144)}$$

So: $\quad r = 13$

$$\tan\theta = \tfrac{12}{5} \Rightarrow \theta = 1.18^{\text{c}} \text{ (3 s.f.)}$$

So: $\quad z = 13(\cos 1.18 + \text{i}\sin 1.18)$

Multiplying complex numbers in modulus–argument form

If $\quad z_1 = r_1(\cos\theta_1 + \text{i}\sin\theta_1)$

and $\quad z_2 = r_2(\cos\theta_2 + \text{i}\sin\theta_2)$

then: $z_1 z_2 = (r_1\cos\theta_1 + \text{i}\,r_1\sin\theta_1)(r_2\cos\theta_2 + \text{i}\,r_2\sin\theta_2)$

$$= r_1 r_2\cos\theta_1\cos\theta_2 + \text{i}\,r_1 r_2\cos\theta_1\sin\theta_2 + \text{i}\,r_1 r_2\sin\theta_1\cos\theta_2 - r_1 r_2\sin\theta_1\sin\theta_2$$

$$= r_1 r_2(\cos\theta_1\cos\theta_2 - \sin\theta_1\sin\theta_2) + \text{i}\,r_1 r_2(\sin\theta_1\cos\theta_2 + \cos\theta_1\sin\theta_2)$$

$$= r_1 r_2\cos(\theta_1 + \theta_2) + \text{i}\,r_1 r_2\sin(\theta_1 + \theta_2)$$

$$= r_1 r_2[\cos(\theta_1 + \theta_2) + \text{i}\sin(\theta_1 + \theta_2)]$$

You can picture what happens geometrically if you draw an Argand diagram:
A_1 represents the number z_1
A_2 represents the number z_2
B represents the product $z_1 z_2$

You can see that the length OB is $r_1 \times r_2$.

You can also see that

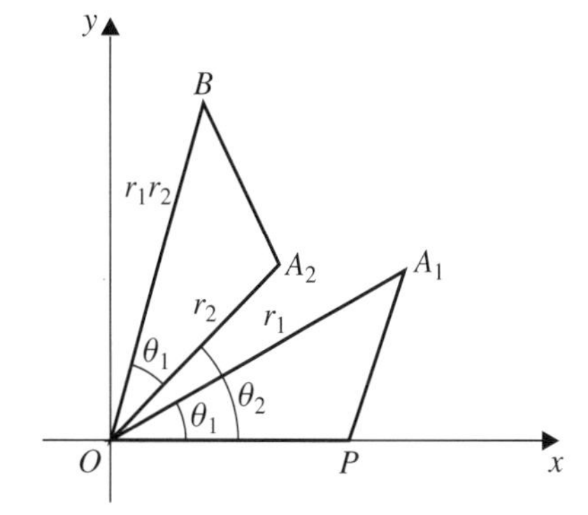

since $\quad \angle A_1OP = \theta_1 = \arg z_1$

$\quad \angle A_2OP = \theta_2 = \arg z_2$

and $\quad \angle BOP = \theta_1 + \theta_2 = \arg(z_1 z_2)$

then $\quad \angle BOA_2 = \theta_1$

If, in addition, the point P has coordinates $(1,0)$, that is P represents the real number 1, then since $\angle BOA_2 = \angle A_1OP (= \theta_1)$ and since

$$\frac{OB}{OA_2} = \frac{r_1 r_2}{r_2} = \frac{r_1}{1}$$

and
$$\frac{OA_1}{OP} = \frac{r_1}{1} = \frac{OB}{OA_2}$$

the triangles OBA_2 and OA_1P are similar.

So when you multiply two complex numbers, you *multiply* their moduli and you *add* their arguments.

Dividing complex numbers in modulus–argument form

Similarly, if

$$z_1 = r_1(\cos\theta_1 + \mathrm{i}\sin\theta_1)$$

and
$$z_2 = r_2(\cos\theta_2 + \mathrm{i}\sin\theta_2)$$

then
$$\begin{aligned}
\frac{z_1}{z_2} &= \frac{r_1\cos\theta_1 + \mathrm{i}\,r_1\sin\theta_1}{r_2\cos\theta_2 + \mathrm{i}\,r_2\sin\theta_2} \\
&= \frac{(r_1\cos\theta_1 + \mathrm{i}\,r_1\sin\theta_1)(r_2\cos\theta_2 - \mathrm{i}\,r_2\sin\theta_2)}{(r_2\cos\theta_2 + \mathrm{i}\,r_2\sin\theta_2)(r_2\cos\theta_2 - \mathrm{i}\,r_2\sin\theta_2)} \\
&= \frac{r_1r_2(\cos\theta_1\cos\theta_2 + \sin\theta_1\sin\theta_2) + \mathrm{i}\,r_1r_2(\sin\theta_1\cos\theta_2 - \cos\theta_1\sin\theta_2)}{r_2^2(\cos^2\theta_2 + \sin^2\theta_2)} \\
&= \frac{r_1}{r_2}[\cos(\theta_1 - \theta_2) + \mathrm{i}\sin(\theta_1 - \theta_2)]
\end{aligned}$$

So when you divide complex numbers you *divide* the modulus of the numerator by the modulus of the denominator and you *subtract* the argument of the denominator from the argument of the numerator.

You should learn these results:

- $|z_1 z_2| = |z_1||z_2|$
- $\left|\frac{z_1}{z_2}\right| = \frac{|z_1|}{|z_2|}$
- $\mathbf{arg}\,(z_1 z_2) = \mathbf{arg}\,z_1 + \mathbf{arg}\,z_2$
- $\mathbf{arg}\,\frac{z_1}{z_2} = \mathbf{arg}\,z_1 - \mathbf{arg}\,z_2$

Example 11
Given that

$$z_1 = 9\left(\cos\frac{7\pi}{12} + \mathrm{i}\sin\frac{7\pi}{12}\right)$$

and
$$z_2 = 5\left(\cos\frac{2\pi}{3} + \mathrm{i}\sin\frac{2\pi}{3}\right)$$

find z_1z_2 in the form $a + \mathrm{i}b$, $a, b \in \mathbb{R}$.

$$\begin{aligned}
z_1z_2 &= 9 \times 5\left[\cos\left(\frac{7\pi}{12} + \frac{2\pi}{3}\right) + \mathrm{i}\sin\left(\frac{7\pi}{12} + \frac{2\pi}{3}\right)\right] \\
&= 45\left[\cos\frac{15\pi}{12} + \mathrm{i}\sin\frac{15\pi}{12}\right] \\
&= 45\left[\cos\frac{5\pi}{4} + \mathrm{i}\sin\frac{5\pi}{4}\right] \\
&= 45\left[\cos\left(\frac{-3\pi}{4}\right) + \mathrm{i}\sin\left(\frac{-3\pi}{4}\right)\right] \\
&= 45\left[\cos\frac{3\pi}{4} - \mathrm{i}\sin\frac{3\pi}{4}\right] \\
&= 45\left(-\frac{1}{\sqrt{2}} - \frac{\mathrm{i}}{\sqrt{2}}\right) = -\frac{45\sqrt{2}}{2}(1 + \mathrm{i})
\end{aligned}$$

Exercise 3C

1 Write the following in the form $r(\cos\theta + \mathrm{i}\sin\theta)$, $-\pi < \theta \leqslant \pi$, giving θ either as a multiple of π or in radians to 3 significant figures.

(a) $5\mathrm{i}$ (b) 7 (c) $-3\mathrm{i}$

(d) -6 (e) $1 + \mathrm{i}\sqrt{3}$ (f) $3\sqrt{3} - 3\mathrm{i}$

(g) $-3 + 4\mathrm{i}$ (h) $1 - \mathrm{i}$ (i) $6 - 8\mathrm{i}$

(j) $\dfrac{2}{1 - \mathrm{i}\sqrt{3}}$ (k) $\dfrac{8}{\sqrt{3} - \mathrm{i}}$ (l) $\dfrac{3 - 2\mathrm{i}}{1 + 4\mathrm{i}}$

2 Write the following in the form $a + \mathrm{i}b$, $a, b \in \mathbb{R}$:

(a) $3\left(\cos\frac{\pi}{3} + \mathrm{i}\sin\frac{\pi}{3}\right)$ (b) $-5\left(\cos\frac{\pi}{4} - \mathrm{i}\sin\frac{\pi}{4}\right)$

(c) $6\left[\cos\left(-\frac{\pi}{6}\right) + \mathrm{i}\sin\left(-\frac{\pi}{6}\right)\right]$ (d) $-4\left(\cos\frac{3\pi}{2} + \mathrm{i}\sin\frac{3\pi}{2}\right)$

(e) $2\left(\cos\frac{2\pi}{7} + \mathrm{i}\sin\frac{2\pi}{7}\right) \times 5\left(\cos\frac{5\pi}{7} + \mathrm{i}\sin\frac{5\pi}{7}\right)$

(f) $\left[3\left(\cos\frac{7\pi}{12} + \mathrm{i}\sin\frac{7\pi}{12}\right)\right]^2$

(g) $\dfrac{7\left(\cos\frac{\pi}{2} + \mathrm{i}\sin\frac{\pi}{2}\right)}{3\left(\cos\frac{\pi}{4} + \mathrm{i}\sin\frac{\pi}{4}\right)}$

(h) $\dfrac{6\left(\cos\frac{\pi}{4} + \mathrm{i}\sin\frac{\pi}{4}\right)}{2\left(\cos\frac{\pi}{2} - \mathrm{i}\sin\frac{\pi}{2}\right)}$

(i) $\left[2\left(\cos\frac{5\pi}{18} + \mathrm{i}\sin\frac{5\pi}{18}\right)\right]^3$

(j) $\dfrac{\left[2\left(\cos\frac{\pi}{4} + \mathrm{i}\sin\frac{\pi}{4}\right)\right]^2}{3\left(\cos\frac{\pi}{3} + \mathrm{i}\sin\frac{\pi}{3}\right)}$

3 Simplify without the use of a calculator

$$\frac{\left(\cos\frac{\pi}{7} - \mathrm{i}\sin\frac{\pi}{7}\right)^3}{\left(\cos\frac{\pi}{7} + \mathrm{i}\sin\frac{\pi}{7}\right)^4}$$

[E]

3.6 Condition for two complex numbers to be equal

If you know that

$$a + \mathrm{i}b = c + \mathrm{i}d, \text{ where } a, b, c, d \in \mathbb{R}$$

then: $a - c = \mathrm{i}(d - b)$

So $(a - c)^2 = \mathrm{i}^2(d - b)^2$

that is: $(a - c)^2 = -(d - b)^2$

Now $(a - c)^2 \geqslant 0$ since when you square a real number the answer cannot be negative. Likewise $(d - b)^2 \geqslant 0$

So: $$-(d - b)^2 \leqslant 0$$

But if $(a - c)^2$ is positive and $-(d - b)^2$ is negative then $(a - c)^2$ cannot equal $-(d - b)^2$. So $(a - c)^2 = 0$ and $-(d - b)^2 = 0$.

Now $(a - c)^2 = 0 \Rightarrow a = c$

and $-(d - b)^2 = 0 \Rightarrow (d - b)^2 = 0 \Rightarrow d = b$

So $$a + \mathrm{i}b = c + \mathrm{i}d$$

$$\Rightarrow \quad a = c \quad \text{and} \quad b = d$$

■ **That is, if two complex numbers are equal, their real parts are equal and their imaginary parts are equal.**

Example 12

Find the square roots of $15 + 8\text{i}$.

If $a + \text{i}b$ is a square root of $15 + 8\text{i}$ then

$$(a + \text{i}b)^2 = 15 + 8\text{i}$$

So:
$$a^2 + 2ab\text{i} - b^2 = 15 + 8\text{i}$$
$$(a^2 - b^2) + 2ab\text{i} = 15 + 8\text{i}$$

If two complex numbers are equal then their real parts are equal and their imaginary parts are equal.

So: $$a^2 - b^2 = 15 \quad (1)$$

and $$2ab = 8 \quad (2)$$

From (2) $$a = \frac{4}{b}$$

Substituting in (1) gives:

$$\frac{16}{b^2} - b^2 = 15$$
$$16 - b^4 = 15b^2$$
$$b^4 + 15b^2 - 16 = 0$$
$$(b^2 + 16)(b^2 - 1) = 0$$
$$b^2 = -16 \quad \text{or} \quad b^2 = 1$$

Since $b \in \mathbb{R}$, b^2 cannot be -16.

So: $b^2 = 1 \Rightarrow b = \pm 1$

Thus $a = \pm 4$

So the square roots are $4 + \text{i}$ and $-4 - \text{i}$.

Example 13

Find the real numbers x and y if

$$3x - 2y + 3\text{i} = 4 + (x + y)\text{i}$$

If $3x - 2y + 3\text{i} = 4 + (x + y)\text{i}$ then equating real and imaginary parts gives:

$$3x - 2y = 4 \quad (1)$$
$$3 = x + y \quad (2)$$

From (2): $$y = 3 - x$$

Substituting into (1) gives:

$$3x - 2(3 - x) = 4$$
$$3x - 6 + 2x = 4$$
$$5x = 10 \Rightarrow x = 2$$

Then $$y = 3 - 2 = 1$$

3.7 Conjugate complex roots of a polynomial equation with real coefficients

In section 3.1 you saw that the roots of the quadratic equation $x^2 + 2x + 5 = 0$ are the complex numbers $-1 + 2\text{i}$ and $-1 - 2\text{i}$.

Notice further that the complex numbers $-1 + 2\text{i}$ and $-1 - 2\text{i}$ are **complex conjugates** (see section 3.2). In fact for any quadratic equation $ax^2 + bx + c = 0$, where a, b, c are real and $b^2 - 4ac < 0$, the roots are $\dfrac{-b + \sqrt{(b^2 - 4ac)}}{2a}$ and $\dfrac{-b - \sqrt{(b^2 - 4ac)}}{2a}$.

These roots are $p + q\text{i}$ and $p - q\text{i}$, where $p = -\dfrac{b}{2a}$ and $q = \dfrac{\sqrt{(4ac - b^2)}}{2a}$ and are therefore complex conjugates.

This result can be generalised for any polynomial equation whose coefficients are real. If the polynomial equation $\text{f}(z) = 0$, with real coefficients, has a root $a + b\text{i}$, where $a, b \in \mathbb{R}$, then the conjugate $a - b\text{i}$ is also a root of $\text{f}(z) = 0$.

You are not expected to prove this important result but you are expected to memorise and use it.

Example 14

Prove that i is a root of the equation $\text{g}(z) = 0$, where $\text{g}(z) = z^3 - 3z^2 + z - 3$. Find the other roots of this equation.

Putting $z = \text{i}$ gives $\quad \text{g(i)} = \text{i}^3 - 3\text{i}^2 + \text{i} - 3$

Remember that $\text{i}^3 = \text{i}^2 \times \text{i} = (-1)\text{i} = -\text{i}$ and $\text{i}^2 = -1$.

Hence $\quad \text{g(i)} = -\text{i} + 3 + \text{i} - 3 = 0$

which proves that i is a root of the equation $\text{g}(z) = 0$.

The conjugate of i is $-\text{i}$ and therefore $-\text{i}$ is also a root of the equation $\text{g}(z) = 0$. You can easily prove this since

$$(-\text{i})^3 - 3(-\text{i})^2 + (-\text{i}) - 3 = -\text{i}^3 - 3\text{i}^2 - \text{i} - 3 = \text{i} + 3 - \text{i} - 3 = 0$$

Two factors of $\text{g}(z)$ are $(z - \text{i})$ and $(z + \text{i})$ and so

$$(z + \text{i})(z - \text{i}) = z^2 - \text{i}^2 = z^2 + 1$$

is a factor of $\text{g}(z)$. By factorising, then

$$z^3 - 3z^2 + z - 3 = (z^2 + 1)(z - 3)$$

which can be checked by multiplying out the brackets or by long division. The third root of the equation $\text{g}(z) = 0$ is therefore 3.

The roots of $\text{g}(z) = 0$ are $\pm\text{i}$, 3.

Example 15

Obtain a quadratic function $f(z) = z^2 + az + b$, where $a, b \in \mathbb{R}$, such that $f(-1 - 2i) = 0$. [E]

If $f(-1 - 2i) = 0$ then $f(-1 + 2i) = 0$.

So: $z - (-1 - 2i) = z + 1 + 2i$
and: $z - (-1 + 2i) = z + 1 - 2i$

are both factors of f.

Thus:
$$(z + 1 + 2i)(z + 1 - 2i) \equiv z^2 + az + b$$
$$z^2 + z - 2iz + z + 1 - 2i + 2iz + 2i + 4 \equiv z^2 + az + b$$
$$z^2 + 2z + 5 \equiv z^2 + az + b$$

That is, $a = 2, \quad b = 5$

and the quadratic function is $z^2 + 2z + 5$.

Example 16

The equation $x^4 - 4x^3 + 3x^2 + 2x - 6 = 0$ has a root $1 - i$. Find the three other roots. [E]

If $1 - i$ is a root then $1 + i$ is a root.

So: $x - (1 - i) = x - 1 + i$

is a factor of

$$x^4 - 4x^3 + 3x^2 + 2x - 6$$

and

$$x - (1 + i) = x - 1 - i$$

is also a factor.

$$(x - 1 + i)(x - 1 - i) = x^2 - x - ix - x + 1 + i + ix - i + 1$$
$$= x^2 - 2x + 2$$

Divide $x^4 - 4x^3 + 3x^2 + 2x - 6$ by $x^2 - 2x + 2$:

$$\begin{array}{r}
x^2 - 2x \quad - 3 \\
x^2 - 2x + 2 \,\big)\, \overline{x^4 - 4x^3 + 3x^2 + 2x - 6} \\
\underline{x^4 - 2x^3 + 2x^2} \\
-2x^3 + x^2 + 2x \\
\underline{-2x^3 + 4x^2 - 4x} \\
-3x^2 + 6x - 6 \\
\underline{-3x^2 + 6x - 6}
\end{array}$$

This means that $x^2 - 2x - 3$ and $x^2 - 2x + 2$ are the quadratic factors of $x^4 - 4x^3 + 3x^2 + 2x - 6$.

Now $x^2 - 2x - 3 = (x - 3)(x + 1)$

which means 3 and -1 are roots of the given quartic equation.

So the three roots of the given equation, in addition to $1 - i$, are -1, 3 and $1 + i$.

Exercise 3D

1 Find the square roots of:

(a) $5 + 12\text{i}$ (b) $7 - 24\text{i}$ (c) $3 - 4\text{i}$

(d) -20i (e) $1 - (4\sqrt{3})\text{i}$

2 Find the real numbers x and y given that:

(a) $x + 4y + xy\text{i} = 12 - 16\text{i}$

(b) $2x + (x - 2y)\text{i} = 18 - y - \text{i}$

(c) $3x + 2x\text{i} = 7 + 2y + (12 + 5y)\text{i}$

(d) $x - 7y + 8x\text{i} = 6y + (6y - 100)\text{i}$

(e) $2x - y + (y - 4)\text{i} = 0$

3 Given that $(1 + 5\text{i})A - 2B = 3 + 7\text{i}$, find A and B if:

(a) A and B are real,

(b) A and B are conjugate complex numbers.

4 Given that $x, y \in \mathbb{R}$ and

$$(x + \text{i}y)(2 + \text{i}) = 3 - \text{i}$$

find x and y.

5 Given that $p, q \in \mathbb{R}$ find p and q where:

(a) $p + q + \text{i}(p - q) = 4 + 2\text{i}$

(b) $2(p + \text{i}q) = q - \text{i}p - 2(1 - \text{i})$

6 Solve for real x and real y the equation

$$(x + \text{i}y)(3 + 4\text{i}) = 3 - 4\text{i}$$ [E]

7 Find the real numbers x and y given that

$$\frac{1}{x + \text{i}y} = 2 - 3\text{i}$$ [E]

8 Given that

$$\frac{1}{x + \text{i}y} + \frac{1}{1 + 2\text{i}} = 1$$

where x and y are real, find x and y. [E]

9 Given that $(a - b\text{i})^2 = -4$, where $a, b \in \mathbb{R}$, find the values of a and b.

10 Given that $z = 5 - 12\text{i}$ express $\frac{1}{z}$ and $z^{\frac{1}{2}}$ in the form $a + \text{i}b$ where $a, b \in \mathbb{R}$.

11 Solve for real values of x and y the equation

$$\frac{x}{1+i} - \frac{y}{2-i} = \frac{1-5i}{3-2i}$$

12 $z_1 = 2 - 3i$, $z_2 = 5 + 4i$.

(a) Express $\frac{z_1 z_2}{z_1 + z_2}$ in the form $A + Bi$ where A and B are real.

(b) Find the real numbers m and n such that

$$mz_1 + nz_2 = 11 + 18i$$

13 Given that the real and imaginary parts of the complex number $z = x + iy$ satisfy the equation

$$(2 - i)x - (1 + 3i)y - 7 = 0$$

find x and y.
State the values of

(a) $|z|$ (b) $\arg z$. [E]

14 (a) The complex number z and its conjugate z^* are given by $z = p + iq$ and $z^* = p - iq$ where p and q are real.
Given that $p = 5$ and $q = 8$

(i) display z and z^* on an Argand diagram

(ii) show that $|z + 2z^*| = 17$

(iii) calculate, in radians to 2 decimal places, the argument of $z + 2z^*$.

(b) Given that

$$ww^* = 5$$

$$\frac{w}{w^*} = \tfrac{1}{5}(-3 + 4i)$$

express w in the form $m + ni$, stating the possible values of the real numbers m and n. [E]

15 Find the values of the real numbers A and B given that

$$(5 + 6i)(3 - 2i) = A + Bi$$

Deduce that

$$(5 - 6i)(3 + 2i) = A - Bi$$

Using your values of A and B deduce that

$$(5^2 + 6^2)(3^2 + 2^2) = 27^2 + 8^2$$

16 Find the values of z, where $z \in \mathbb{C}$, for which

$$z^4 - 1 = 0$$

17 One root of the equation $2z^3 - 9z^2 + 30z - 13 = 0$ is $2 + 3i$. Find the other two roots.

18 One root of the equation $2z^3 - 5z^2 + 12z - 5 = 0$ is $1 - 2i$. Find the other two roots.

19 One root of the equation $z^4 + 3z^3 + 12z - 16 = 0$ is $2i$. Find the other three roots.

20 Find the values of z, where $z \in \mathbb{C}$, for which

$$z^4 + 5z^2 + 4 = 0$$

21 Prove that $z^2 - z + 2$ is a factor of $f(z)$, where

$$f(z) \equiv z^3 - 3z^2 + 4z - 4$$

Hence find the values of z, where $z \in \mathbb{C}$, for which $f(z) = 0$.

22 Given that a and b are real constants, prove that ai is a root of the equation

$$z^3 - bz^2 + a^2z - a^2b = 0$$

Find the other roots of the equation in terms of a and b.

SUMMARY OF KEY POINTS

1. $\sqrt{(-1)} = i$
2. A number of the form bi, where b is real, is called a pure imaginary number.
3. A number of the form $a + bi$, where $a, b \in \mathbb{R}$, is called a complex number.
4. If $z = x + iy$ then the complex conjugate of z is $z^* = x - iy$.
5. Any complex number can be represented by either a point or a vector on an Argand diagram.
6. If $z = x + iy$ then the modulus of z is
$$|z| = \sqrt{(x^2 + y^2)}$$
7. If $z = x + iy$ then $\arg z$ is the principal value of the argument of z.

8 If $a + \mathrm{i}b = c + \mathrm{i}d$, where a, b, c, $d \in \mathbb{R}$, then $a = c$ and $b = d$.

9 If
$$z = x + \mathrm{i}y = r(\cos\theta + \mathrm{i}\sin\theta)$$
where $-\pi < \theta \leqslant \pi$ and θ is the angle that the vector representing z on an Argand diagram makes with the positive x-axis, then
$$x = r\cos\theta \quad y = r\sin\theta$$
$$r = \sqrt{(x^2 + y^2)}$$

10 If $z_1 = r_1(\cos\theta_1 + \mathrm{i}\sin\theta_1)$ and $z_2 = r_2(\cos\theta_2 + \mathrm{i}\sin\theta_2)$ then
$$z_1 z_2 = r_1 r_2[\cos(\theta_1 + \theta_2) + \mathrm{i}\sin(\theta_1 + \theta_2)]$$
$$\frac{z_1}{z_2} = \frac{r_1}{r_2}[\cos(\theta_1 - \theta_2) + \mathrm{i}\sin(\theta_1 - \theta_2)]$$
$$|z_1 z_2| = |z_1||z_2|$$
$$\left|\frac{z_1}{z_2}\right| = \frac{|z_1|}{|z_2|}$$
$$\arg(z_1 z_2) = \arg z_1 + \arg z_2$$
$$\arg\frac{z_1}{z_2} = \arg z_1 - \arg z_2$$

11 If the polynomial equation $\mathrm{f}(z) = 0$, with real coefficients, has a root $a + b\mathrm{i}$, where $a, b \in \mathbb{R}$, then the conjugate $a - b\mathrm{i}$ is also a root of the equation $\mathrm{f}(z) = 0$.

Numerical solution of equations

4

Chapter 4 of Book C3 describes how you can detect an interval in which a root of the equation $f(x) = 0$ lies. You can do this by using the fact that, in general, a root lies in the interval $[a, b]$ if $f(a) < 0$ and $f(b) > 0$ or vice versa.

Chapter 4 of Book C3 also introduced you to some iterative procedures for finding the root of an equation to whatever degree of accuracy you require. In this chapter you will learn three other procedures for finding approximations to the roots of equations:

(a) linear interpolation
(b) interval bisection
(c) the Newton–Raphson process.

4.1 Review of iterative procedures

The technique of iteration gives a sequence of approximations. Usually this sequence converges to a root of the equation. However, sometimes the sequence diverges and takes you further and further away from the root.

Here is an example where the iterative procedure is given and it converges to a root.

Example 1
Show that the equation $f(x) = 0$, where $f(x) \equiv 4x - \sec^2 x$, has a root in the interval $[0.2, 0.3]$.

Use the iterative procedure

$$x_{n+1} = \tfrac{1}{4}\sec^2 x_n \text{ with } x_1 = 0.2$$

to find this root correct to 4 decimal places.

$f(0.2) = -0.241\,09\ldots$, $f(0.3) = 0.104\,311\ldots$ so a root lies in the interval [0.2, 0.3].

$$x_{n+1} = \tfrac{1}{4}\sec^2 x_n;\ x_1 = 0.2$$

So, using a calculator you obtain:

$$x_2 = 0.260\,272\,8\ldots$$

$$x_3 = 0.267\,730\,7\ldots$$

$$x_4 = 0.268\,812\,3\ldots$$

$$x_5 = 0.268\,972\,3\ldots$$

$$x_6 = 0.268\,996\,0\ldots$$

Since both x_5 and x_6 give the root as 0.2690 to 4 decimal places it is now necessary to test this to see if the root is 0.2690 to this degree of accuracy.

To do this you need to go either side of 0.2690. So you consider the interval [0.268 95, 0.269 05]:

$$f(0.268\,95) = -0.000\,170\,8\ldots$$
$$f(0.269\,05) = 0.000\,169\,85\ldots$$

Since there is a sign change, the root does indeed lie in the interval [0.268 95, 0.269 05] and so the root is 0.2690 correct to 4 d.p.

4.2 Linear interpolation

If you want to find a root of $f(x) = 0$, another method of finding the approximate value of the root is to first consider the graph of $y = f(x)$.

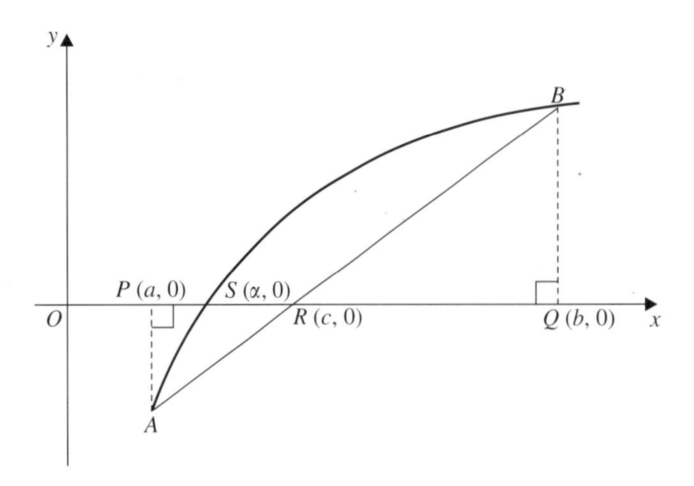

From the graph you can see that since the curve cuts the x-axis at $S(\alpha, 0)$, a root of the equation $f(x) = 0$ is α, and it lies in the interval $[a, b]$. To find an approximation to α you join A and B with a straight line and use the point where this straight line crosses the x-axis as a first approximation to the root. So in this case you try to find c and use this as an approximation to α.

To find c, use the fact that triangles PAR and QBR are similar ($\angle APR = \angle BQR = 90^\circ$ and $\angle ARP = \angle BRQ$ since they are vertically opposite). You can then see that

$$\frac{PR}{AP} = \frac{QR}{BQ}$$

From this you can work out the value of c.

You can repeat this process to get a closer approximation to α, and go on repeating it until you have a value for the root to the desired degree of accuracy, as the next example shows.

Example 2

Show that the equation $x^3 + 5x - 10 = 0$ has a root in the interval [1, 2]. Using linear interpolation, find this root to 1 decimal place.

Let $f(x) \equiv x^3 + 5x - 10$. Then:

$$f(1) = 1 + 5 - 10 = -4$$

$$f(2) = 8 + 10 - 10 = 8$$

Since there is a sign change, the equation $x^3 + 5x - 10 = 0$ has a root in [1, 2].

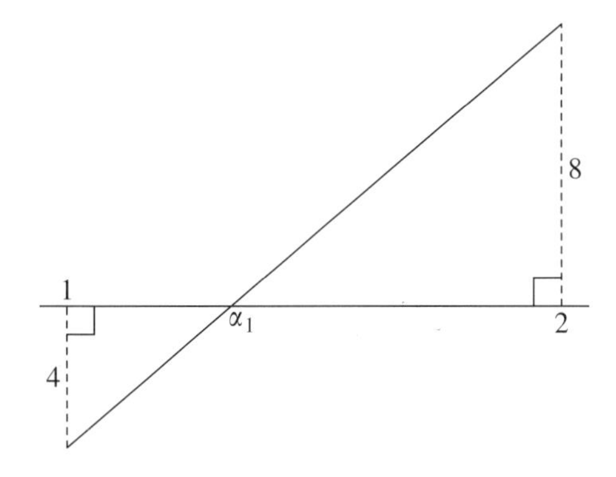

Using similar triangles:

$$\frac{2-\alpha_1}{\alpha_1-1}=\frac{8}{4}$$

So:
$$2-\alpha_1=2\alpha_1-2$$

$$4=3\alpha_1$$

$$\alpha_1=\frac{4}{3}=1.333\ldots$$

$f(1.333\ldots)=-0.962\,96\ldots$ Since $f(2)>0$ and $f(1.333\ldots)<0$, the root lies in the interval $[1.333\ldots, 2]$.

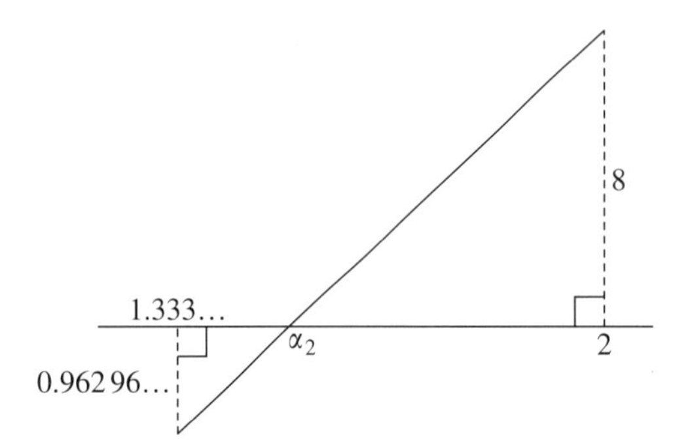

By similar triangles:

$$\frac{2-\alpha_2}{\alpha_2-1.333\ldots}=\frac{8}{0.962\,96\ldots}$$

$$\alpha_2=1.404\,95\ldots$$

$$f(1.404\,95\ldots)=-0.2019\ldots$$

So the root lies in the interval $[1.404\,95\ldots, 2]$.

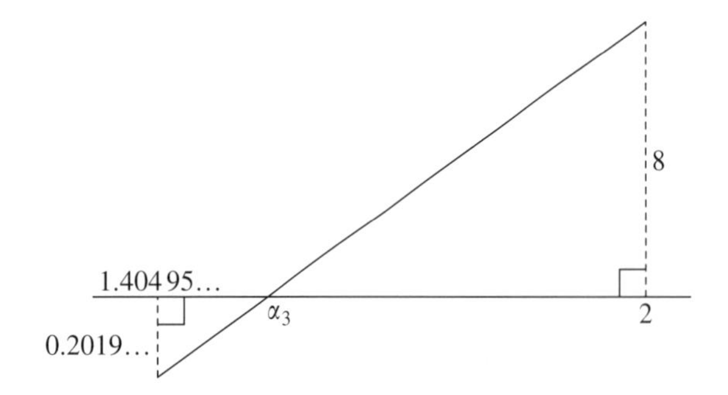

$$\frac{2-\alpha_3}{\alpha_3-1.404\,95\ldots}=\frac{8}{0.2019\ldots}$$

$$\alpha_3=1.4196\ldots$$

Two successive approximations give the root as 1.4 to 1 decimal place, so you can now test either side of 1.4 by considering the interval [1.35, 1.45]:

$$f(1.35) = -0.789\,625$$

$$f(1.45) = 0.298\,62\ldots$$

Since $f(1.35) < 0$ and $f(1.45) > 0$, the root lies in [1.35, 1.45] and so $\alpha = 1.4$ (1 d.p.).

4.3 Interval bisection

Another method of finding an approximation to a root of the equation $f(x) = 0$ is to find an interval $[a, b]$ in which the root lies and then take the mid-point $\frac{a+b}{2}$ of this interval as a first approximation to the root.

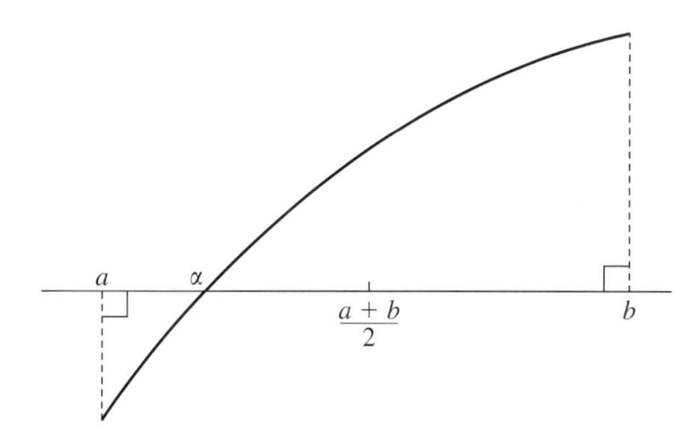

When you have done this it is obvious that the root will then lie either in the left-hand interval $\left[a, \frac{a+b}{2}\right]$ or in the right-hand interval $\left[\frac{a+b}{2}, b\right]$.

You decide which one it is (in the diagram it is the left-hand interval), and then choose the mid-point of this interval as a second approximation to the root. You then repeat the process until you obtain the root to the accuracy required.

Example 3

Use interval bisection to obtain a sequence of approximations to the positive root of the equation $x^2 - 7 = 0$. Hence find $\sqrt{7}$ to 2 decimal places.

Let $f(x) \equiv x^2 - 7$

Since $f(2) = -3 < 0$ and $f(3) = 2 > 0$ the positive root lies in [2, 3]. Take $x_0 = 2$ and $x_1 = 3$. Then:

$$x_2 = \frac{2+3}{2} = 2.5 \text{ and } f(2.5) = -0.75$$

$f(2.5) < 0$ and $f(3) > 0$ so the root lies in [2.5, 3].

So: $$x_3 = \frac{2.5+3}{2} = 2.75 \text{ and } f(2.75) = 0.5625$$

$f(2.75) > 0$ and $f(2.5) < 0$ so the root lies in [2.5, 2.75].

Thus: $$x_4 = \frac{2.5+2.75}{2} = 2.625 \text{ and } f(2.625) = -0.1093\ldots$$

$f(2.625) < 0$ and $f(2.75) > 0$ so the root lies in [2.625, 2.75].

Thus: $$x_5 = \frac{2.625+2.75}{2} = 2.6875 \text{ and } f(2.6875) = 0.222\,656\,25$$

$f(2.6875) > 0$ and $f(2.625) < 0$ so the root lies in [2.625, 2.6875].

Thus: $$x_6 = \frac{2.625+2.6875}{2} = 2.656\,25$$

and $$f(2.656\,25) = 0.055\,664\ldots$$

$f(2.656\,25) > 0$ and $f(2.625) < 0$ so the root lies in [2.625, 2.656 25].

Thus: $$x_7 = \frac{2.625+2.656\,25}{2} = 2.640\,625$$

and $$f(2.640\,625) = -0.027\,09\ldots$$

$f(2.640\,625) < 0$ and $f(2.656\,25) > 0$
so the root lies in [2.640 625, 2.656 25].

Thus $$x_8 = \frac{2.640\,625+2.656\,25}{2} = 2.648\,437\,5$$

and $$f(2.648\,437\,5) = 0.014\,22\ldots$$

$f(2.648\,437\,5) > 0$ and $f(2.640\,625) < 0$
so the root lies in [2.640 625, 2.648 437 5].

So: $$x_9 = \frac{2.640\,625 + 2.648\,437\,5}{2} = 2.644\,531\,25$$

and $$f(2.644\,531\,25) = -0.006\,45\ldots$$

$f(2.644\,531\,25) < 0$ and $f(2.648\,437\,5) > 0$
so the root lies in [2.644 531 25, 2.648 437 5]

So: $$x_{10} = \frac{2.644\,531\,25 + 2.648\,437\,5}{2} = 2.646\,484\,375$$

and $$f(2.646\,484\,375) = 0.003\,87\ldots$$

$f(2.646\,484\,375) > 0$ and $f(2.644\,531\,25) < 0$
so the root lies in [2.644 531 25, 2.646 484 375].

So: $$x_{11} = \frac{2.644\,531\,25 + 2.646\,484\,375}{2} = 2.645\,507\,813$$

and $$f(2.645\,507\,813) = -0.001\,28\ldots$$

The root lies in [2.645 507 813, 2.646 484 375]. Since both ends of the interval when corrected to 2 decimal places are 2.65, the root is 2.65 to 2 decimal places.

If you put these calculations in a table, the working becomes more concise and less prone to errors:

a	b	$\frac{a+b}{2}$
2	3	2.5
2.5	3	2.75
2.5	2.75	2.625
2.625	2.75	2.6875
2.625	2.6875	2.656 25
2.625	2.656 25	2.640 625
2.640 625	2.656 25	2.648 437 5
2.640 625	2.648 437 5	2.644 531 25
2.644 531 25	2.648 437 5	2.646 484 375
2.644 531 25	2.646 484 375	2.645 507 813

The root is 2.65 to 2 decimal places.

As you can see, while this method is very easy to apply, it sometimes takes a long time to get the root to the accuracy that you require.

4.4 The Newton–Raphson process

You can derive another method of finding an approximation to a root of the equation $f(x) = 0$ by considering the graph of $y = f(x)$:

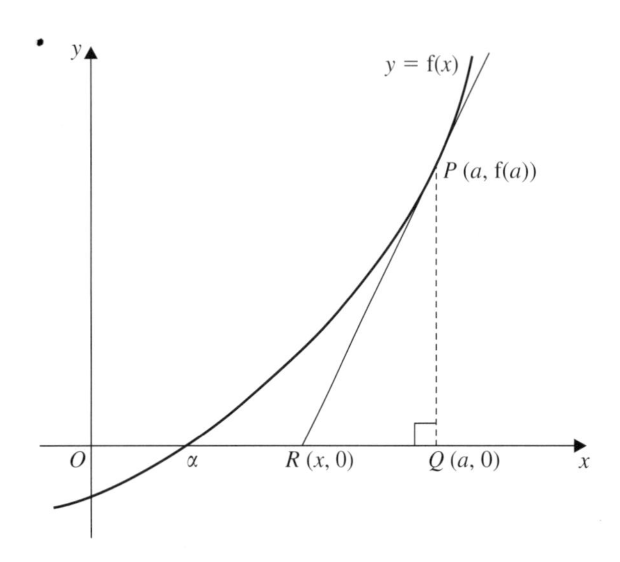

As you can see from the diagram α is a root of $f(x) = 0$. Take a as a first approximation to α. Since the point P on the curve has a as its x-coordinate, draw the tangent to the curve at P. This cuts the x-axis at $R(x, 0)$.

The gradient of the curve $y = f(x)$ is given by $\frac{dy}{dx} = f'(x)$. So the gradient of the curve at P is $f'(a)$. Using the fact that at P, $x = a$, $y = f(a)$ and $\frac{dy}{dx} = f'(a)$, the equation of the tangent at P is

$$y - f(a) = f'(a)(x - a)$$

The tangent cuts the x-axis where $y = 0$, so at R:

$$-f(a) = f'(a)(x - a)$$

That is:
$$x - a = \frac{-f(a)}{f'(a)}$$

or:
$$x = a - \frac{f(a)}{f'(a)}$$

So the x-coordinate of R is $a - \frac{f(a)}{f'(a)}$.

■ **That is, if a is a first approximation to a root of $f(x) = 0$, a better approximation is, in general,**

$$a - \frac{f(a)}{f'(a)}$$

You can now use this value, repeat the process, and obtain a third approximation, which will, in general, be a better approximation than the second, and so on.

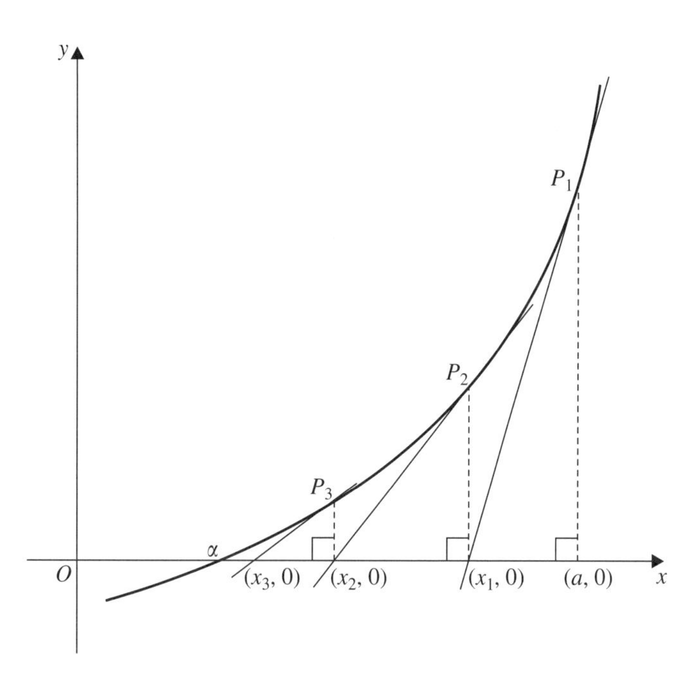

As you can see from the diagram, each successive tangent cuts the x-axis closer and closer to the point where the curve cuts the x-axis. Consequently, the sequence of approximations a, x_1, x_2, $x_3, \ldots$ is getting closer and closer to the root α of the equation $f(x) = 0$.

Although the Newton–Raphson process works well in general, there are occasions on which the sequence of approximations takes you further and further from a root.

For example, you can see in the diagram on the next page that the tangent at $(a, f(a))$ cuts the x-axis at a point which is further from $(\alpha, 0)$ than is $(a, 0)$. Under these circumstances the Newton–Raphson process fails.

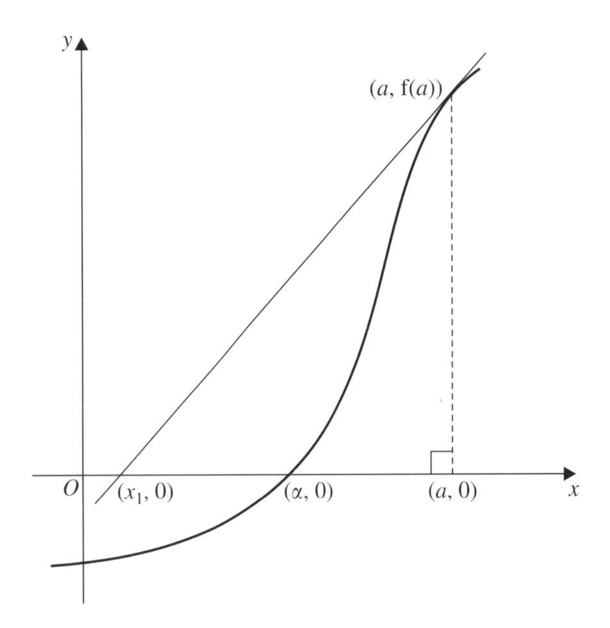

Example 4

Show that a root of the equation $\ln x = 4 - x$ lies between 2.9 and 3.

By taking 2.9 as a first approximation and applying the Newton–Raphson process once to the equation $\ln x - 4 + x = 0$, find a second approximation, giving your answer to 3 significant figures.

Let $f(x) \equiv \ln x - 4 + x$

then: $\quad f(2.9) = -0.0352\ldots$

and $\quad f(3) = 0.0986\ldots$

Since there is a change of sign a root lies between 2.9 and 3.

$$f(x) \equiv \ln x - 4 + x$$

$$f'(x) \equiv \frac{1}{x} + 1$$

$$x_0 = 2.9$$

Using $x_0 - \dfrac{f(x_0)}{f'(x_0)}$ as a better approximation you get

$$x_1 = 2.9 - \frac{\ln 2.9 - 4 + 2.9}{\dfrac{1}{2.9} + 1}$$

$$= 2.9262\ldots$$

$$= 2.93 \text{ (3 s.f.)}$$

Example 5

Show that the equation $4x^2 - 1 - 2\tan x = 0$ has a root α in the interval [–0.4, –0.3]. Use the Newton–Raphson procedure once, starting with –0.3, to find a second approximation to α, giving your answer to 3 decimal places. Show further that this approximation to α is correct to 3 decimal places.

Let $f(x) \equiv 4x^2 - 1 - 2\tan x$

$$f(-0.4) = 0.4855\ldots$$
$$f(-0.3) = -0.0213\ldots$$

Since there is a sign change, a root lies in the interval [–0.4, –0.3].

$$f(x) \equiv 4x^2 - 1 - 2\tan x$$
$$f'(x) \equiv 8x - 2\sec^2 x$$

$$x_0 = -0.3$$
$$x_1 = -0.3 - \frac{4(-0.3)^2 - 1 - 2\tan(-0.3)}{8(-0.3) - 2\sec^2(-0.3)}$$
$$= -0.3046\ldots$$

So the root is –0.305 (3 d.p.).

Consider the interval [–0.3055, –0.3045] to check that the root is correct to 3 decimal places:

$$f(-0.3055) = 0.004\,066\ldots$$
$$f(-0.3045) = -0.000\,571\ldots$$

The sign change indicates that the root is –0.305 correct to 3 decimal places.

Exercise 4A

1 Show that a root of the equation $x^3 - 5x - 3 = 0$ lies in the interval [2, 3]. Use linear interpolation to find this root correct to 2 decimal places.

2 Show that a root of the equation $4 = xe^{2x}$ lies in the interval [0, 1]. Use linear interpolation to find this root correct to 2 decimal places.

3 Show that a root of the equation $4x^3 - 9x^2 + 3 = 0$ lies in the interval [0, 1]. Use linear interpolation to find this root correct to 2 decimal places.

4 Show that a root of the equation $x + 2 = \frac{4}{x}$ lies in the interval [1, 2]. Use linear interpolation to find this root correct to 2 decimal places.

5 Show that the equation $xe^x - 1 = x$ has a root in the interval [0, 1]. Use linear interpolation to find this root correct to 2 decimal places.

6 Show that the equation $4x - 3 - e^{-x}(x^2 + 1) = 0$ has a root α in the interval [0.9, 1]. By using linear interpolation find an approximation x_0 to α. Taking x_0 as a first approximation to α, use the Newton–Raphson process once to find a second approximation to α, giving your answer to 3 decimal places.

7 Verify that the equation $x^3 - x^2 + 8x - 4 = 0$ has a root α in the interval [0.5, 0.6]. Use a sequence of linear interpolations to find α correct to 2 decimal places.

8 Show that the equation $4x^2 - \tan x - 1 = 0$ has a root α in the interval [0.6, 0.7]. Use the iterative procedure

$$x_{n+1} = \tfrac{1}{2}\sqrt{(1 + \tan x_n)}$$

with $x_0 = 0.7$ to find α correct to 3 decimal places.

9 Show that $x^5 + 5x - 1 = 0$ has a root α in the interval [0.1, 0.3]. Taking $x_0 = 0.3$, use the iterative procedure

$$x_{n+1} = \tfrac{1}{5}(1 - x_n^5)$$

to find α correct to 6 decimal places.

10 Show that the largest positive root of the equation $x^3 - 3x^2 - 2 = 0$ lies between 3 and 4. Use a sequence of linear interpolations to estimate this root correct to 2 decimal places.

11 Use interval bisection to find, correct to 2 decimal places, the root of the equation $x^4 + 3x^2 - x - 8 = 0$ that lies in the interval [1, 2].

12 Use interval bisection to find the positive root of the equation $x^2 - 5 = 0$, correct to 2 decimal places.

13 Show that a root of the equation $x^3 - 5x - 3 = 0$ lies in the interval [−2, −1]. Use interval bisection to find this root correct to 2 decimal places.

14 Show that a root of the equation $e^{2x} + 3x = 5$ lies in the interval [0, 1]. Use interval bisection to find this root correct to 3 decimal places.

15 Show that the equation

$$\frac{15}{2+x} = 3 + \sqrt{x}$$

has a root lying in the interval [1, 2]. Use interval bisection to find this root correct to 2 decimal places.

16 Show that the equation $f(x) = 0$ has a root in the given interval. Use the Newton–Raphson process once, starting with x_0, to find a second approximation and give your answer to 2 decimal places.

(a) $f(x) \equiv x^3 - x^2 + 6x - 1$; [0.15, 0.20]; $x_0 = 0.2$

(b) $f(x) \equiv x^3 + 2x^2 + 5x - 27$; [2, 2.2]; $x_0 = 2.1$

(c) $f(x) \equiv x^3 - 30$; [3, 3.2]; $x_0 = 3.2$

(d) $f(x) \equiv \tan x - e^x$; $[0, \frac{5\pi}{12}]$; $x_0 = 1.3$

(e) $f(x) \equiv x^3 + x^2 - 100$; [4, 5]; $x_0 = 4.4$

(f) $f(x) \equiv x + \sin x - 1$; [0.4, 0.6]; $x_0 = 0.54$

(g) $f(x) \equiv \ln x - 1 - \frac{1}{x}$; [3, 4]; $x_0 = 3.6$

17 Show that the equation $x^3 - 12x - 7.2 = 0$ has one positive and two negative roots. Obtain the positive root to 3 significant figures using the Newton–Raphson process. [E]

18 Find, correct to 1 decimal place, the real root of $x^3 + 2x - 1 = 0$ by using the Newton–Raphson process. [E]

19 Using the same axes draw accurate graphs of $y = \ln x$ and $y = 3 - x$ in the interval $1 \leqslant x \leqslant 4$. Deduce that the equation $x + \ln x - 3 = 0$ has a root near 2.2. Clearly showing your method, obtain alternative approximations to the root of the equation

(a) by linear interpolation between $x = 2.2$ and $x = 2.3$

(b) by one application of the Newton–Raphson process using $x = 2.2$ as the initial value. [E]

20 Use the Newton–Raphson process to find the real root of the equation $x^3 + 2x^2 + 4x - 6 = 0$, taking $x = 0.9$ as the first approximation and carrying out one iteration. [E]

21 Show that the equation $x^3 + 3x - 3 = 0$ has only one real root and that it lies between 0.8 and 1. Obtain approximations to the root

(a) by performing one application of the Newton–Raphson process using $x = 0.8$ as the first approximation

(b) by performing two iterations using the procedure defined by

$$x_{n+1} = \frac{(3 - x_n^3)}{3}$$

and starting with $x = 0.8$. [E]

22 Given that $f(x) \equiv 4x - e^x$, show that the equation $f(x) = 0$ has a root α in the interval [0.3, 0.4]. Taking 0.35 as an initial approximation to α, use the Newton–Raphson process twice to obtain two further approximations to α, giving your final answer to 3 decimal places.

23

$$f(x) \equiv x^3 + 3x + 3$$

(a) Show that the equation $f(x) = 0$ has a root α in the interval $[-1, 0]$.

(b) Use linear interpolation once on the interval $[-1, 0]$ to obtain an approximation x_0 to α.

(c) Using your value x_0, use the Newton–Raphson procedure once to find a second approximation x_1 to α, giving your answer to 2 decimal places.

(d) Show that x_1 is correct to 2 decimal places.

24 It is given that

$$f(x) \equiv x - (\sin x + \cos x)^{\frac{1}{2}}, \; 0 \leqslant x \leqslant \tfrac{3}{4}\pi.$$

(a) Show that the equation $f(x) = 0$ has a root lying between 1.1 and 1.2.

(b) Using 1.2 as a first approximation to this root, apply the Newton–Raphson procedure once to obtain a second approximation, giving your answer to 2 decimal places.

SUMMARY OF KEY POINTS

1 In order to find a root of the equation $f(x) = 0$ by iteration, the equation must first be rearranged in the form $x = g(x)$. An iteration formula is then

$$x_{n+1} = g(x_n)$$

2 To find a root of an equation $f(x) = 0$ by linear interpolation, use a straight line to join two points with x-coordinates a and b on the graph of $y = f(x)$ that lie on opposite sides of the x-axis. Take the point where the line cuts the x-axis as a first approximation α_1 to the root α and work out its value by similar triangles. Work out $f(\alpha_1)$ to find out whether the root lies in the interval $[a, \alpha_1]$ or $[\alpha_1, b]$ and repeat the process on the appropriate interval to find a closer approximation to α, etc.

3 Interval bisection: If a root α of the equation $f(x) = 0$ lies in the interval $[a, b]$, the mid-point $\dfrac{a+b}{2}$ is a first approximation to α. Calculate $f(a)$, $f(b)$ and $f\left(\dfrac{a+b}{2}\right)$ to find out whether the root lies in the interval $\left[a, \dfrac{a+b}{2}\right]$ or $\left[\dfrac{a+b}{2}, b\right]$ and then find the mid-point of the appropriate interval to find a closer approximation to α, etc.

4 The Newton–Raphson process:
If a is a first approximation to a root of $f(x) = 0$, then

$$a - \frac{f(a)}{f'(a)}$$

is in general a better approximation.

Review exercise 1

1 Given that $z_1 = 2 + 3\text{i}$ and $z_2 = 2 - \text{i}$,

(a) express $z_1 z_2$ and $\frac{z_1}{z_2}$ in the form $a + b\text{i}$, where $a, b \in \mathbb{R}$

(b) display the complex numbers z_1, z_2, $z_1 z_2$ and $\frac{z_1}{z_2}$ on the same Argand diagram.

2 Given that $\text{f}(x) \equiv x^3 - 4x + 1$, show that the equation $\text{f}(x) = 0$ has a root α in the interval $[1, 2]$.
Find the value of α correct to 2 decimal places.

3 Find the set of values of x for which

$$\frac{2}{x-4} < 5$$

4 Find $\sum_{r=1}^{n} r(r+1)(r+4)$ giving your answer in the form $An(n+1)(n+2)(Bn+C)$ where the constants A, B and C are to be found.

5 Find (a) $\sum_{r=1}^{50} r^2$ (b) $\sum_{r=1}^{33} (3r)^2$ (c) $\sum_{r=1}^{40} (2r-1)^2$

6 In an Argand diagram, the complex numbers $z_1 = x_1 + \text{i}y_1$ and $z_2 = x_2 + \text{i}y_2$ are represented by the points A and B respectively. Prove that the complex number z_3 given by $z_3 = 5z_2 - 4z_1$ and represented by the point C in the Argand diagram lies on the line passing through A and B.

7

$$\text{f}(x) \equiv x^5 - 4x - 1$$

(a) Prove that the equation $\text{f}(x) = 0$ has a root α in the interval $[1, 2]$.

(b) By using the iterative relation

$$u_{n+1} = (4u_n + 1)^{\frac{1}{5}},\ u_0 = 1$$

or otherwise, determine the value of α, correct to two decimal places.

8 Find the set of values of t for which

$$|t - 2| > t^2$$

9 Given that $z = -\frac{1}{2} + \frac{\sqrt{3}}{2}\mathrm{i}$, find the modulus and the argument of $\frac{1}{1+z}$, giving the argument θ in radians, $-\pi < \theta \leqslant \pi$.

10

$$\mathrm{f}(x) \equiv x^3 - 3x^2 - 5$$

(a) Find the positive integer N such that the equation $\mathrm{f}(x) = 0$ has a root α in the interval $[N, N+1]$.
(b) Use linear interpolation on the interval $[N, N+1]$ to find an approximation for α, giving your answer to two decimal places.

11 Sketch the curve with equation $y = |\cos x|$ over the interval $-\pi < x < 2\pi$, showing the coordinates of those points where the curve meets the x-axis. Hence, or otherwise, find the set of values of x for which

$$|\cos x| < \frac{\sqrt{3}}{2},\quad -\pi < x < 2\pi$$

12 The complex number z is such that

$$z - 3z^* = -6 + 4\mathrm{i}$$

Determine z in the form $a + \mathrm{i}b$, where $a, b \in \mathbb{R}$ and display z and z^* on an Argand diagram.

13 Evaluate $\sum_{r=11}^{30} r(3r - 1)$. [E]

14 Find the set of values of x for which

$$\frac{2}{x+2} > \frac{1}{x-3}$$ [E]

15 $$f(x) \equiv x^3 + x^2 - 6$$

(a) Show that the real root α of the equation $f(x) = 0$ lies in the interval $[1, 2]$.

(b) Use linear interpolation on the interval $[1, 2]$ to find a first approximation to α.

(c) Use the Newton–Raphson procedure on $f(x)$ once, starting with your answer to (b), to find another approximation to α, giving your answer to 2 decimal places.

[E]

16 (a) Find, in radians to 2 decimal places, the argument of the complex number $-6 + 5i$.

(b) Given that $(-6 + 5i)(p + qi) = -13 + 21i$, $p, q \in \mathbb{R}$, find the value of p and the value of q.

(c) Using your values of p and q, show that $p + qi$ is a root of the equation $z^2 - 6z + 10 = 0$ and find the other root of this equation. [E]

17 The complex number z satisfies the equation

$$\frac{z - 2}{z + 3i} = \lambda i, \quad \lambda \in \mathbb{R}$$

(a) Show that $z = \dfrac{(2 - 3\lambda)(1 + \lambda i)}{1 + \lambda^2}$

(b) In the case when $\lambda = 1$ find $|z|$ and $\arg z$. [E]

18 Starting with $x = 1.5$, apply the Newton–Raphson procedure once to $f(x) \equiv x^3 - 3$ to obtain a better approximation to the cube root of 3, giving your answer to 3 decimal places. [E]

19 (a) Sketch the graphs of $y = |x - 8|$ and $y = 8x$ using the same pair of axes.

(b) Determine the set of values of x for which $|x - 8| > 8x$.

[E]

20 Show that $\sum_{r=1}^{2n} (2r - 1)^2 = \frac{2}{3}n(16n^2 - 1)$. [E]

21 (a) Show the equation $x^3 - 3x + 1 = 0$ has a root α lying between 1.5 and 1.6.

Given that x_0 is an approximate solution to this equation, a better approximation x_1 is sought using the iterative formula

$$x_1 = \sqrt{\left(\frac{3x_0 - 1}{x_0}\right)}$$

(b) Take 1.51 as a first approximation to α, and apply this iterative formula twice to obtain two further approximations to α. Hence state the value of α as accurately as your working justifies. [E]

22 Find the set of real values of x for which

$$\frac{2x^2 + 6}{x + 6} > 1$$ [E]

23 Express $\frac{2}{4x^2 - 1}$ in partial fractions.

Hence, or otherwise, show that

$$\sum_{r=1}^{n} \frac{2}{4r^2 - 1} = \frac{2n}{2n + 1}$$ [E]

24 The complex number z is given by

$$z = \frac{3 + \mathrm{i}}{2 - \mathrm{i}}$$

(a) Show that $\arg z = 45°$ and find $|z|$.

The complex number z is represented by the point P in an Argand diagram, origin O. The complex number $-5 + k\mathrm{i}$ is represented by the point Q and $\angle POQ$ is 90°.

(b) Find the value of k.

(c) Find the complex number w, represented by the mid-point M of PQ.

(d) Calculate $\arg w$, giving your answer in degrees to 1 decimal place [E]

25 (a) Show that there is a root of the equation $8 \sin x - x = 0$ lying between 2.7 and 2.8.

(b) Taking 2.8 as a first approximation to this root, apply the Newton–Raphson procedure once to $\mathrm{f}(x) \equiv 8 \sin x - x$ to obtain a second approximation, giving your answer to 2 decimal places.

(c) Explain, with justification, whether or not this second approximation is correct to 2 decimal places.

(d) Evaluate $f\left(\frac{5\pi}{2}\right)$, and hence, by sketching suitable graphs, determine the number of roots of the equation $8\sin x = x$ in the range $x > 0$. [E]

26 Given that $1 + 2i$ is a root of the equation

$$x^4 - 4x^3 - 6x^2 + 20x - 75 = 0$$

find the other three roots.

Plot the points representing all four of these roots on an Argand diagram. Hence, or otherwise, show that these points are the vertices of a rhombus with sides of length $2\sqrt{5}$. [E]

27 Using the same axes, sketch the curves with equations

$$y = \frac{1}{x} \quad \text{and} \quad y = \frac{x}{x+2}$$

State the equations of any asymptotes, the coordinates of any points of intersection with the axes and the coordinates of any points of intersection of the two curves.

Hence, or otherwise, find the set of values of x for which

$$\frac{1}{x} > \frac{x}{x+2}$$

[E]

28 Given that $f(r) = \frac{1}{r^2}$, show that

$$f(r) - f(r+1) = \frac{2r+1}{r^2(r+1)^2}$$

and hence find $\sum_{r=1}^{n} \frac{2r+1}{r^2(r+1)^2}$. [E]

29 The tangent at $P(x_n, x_n^2 - 2)$, where $x_n > 0$, to the curve with equation $y = x^2 - 2$ meets the x-axis at the point $Q(x_{n+1}, 0)$. Show that

$$x_{n+1} = \frac{x_n^2 + 2}{2x_n}$$

This relationship between x_{n+1} and x_n is used, starting with $x_1 = 2$, to find successive approximations for the positive root of the equation $x^2 - 2 = 0$. Find x_2 and x_3 as fractions and show that $x_4 = \frac{577}{408}$.

Find the error, to 1 significant figure, in using $\frac{577}{408}$ as an approximation to $\sqrt{2}$. [E]

30 Given that $3 - 5\text{i}$ is a solution of the equation $\text{P}(z) = 0$, where

$$\text{P}(z) \equiv z^4 - 8z^3 + 43z^2 - 50z - 102$$

factorise the polynomial $\text{P}(z)$ into linear and quadratic factors with real coefficients.

Find the other three solutions of the equation $\text{P}(z) = 0$. [E]

31 Find the complete set of values of x for which

$$\frac{3x}{x-1} > x$$ [E]

32 (a) Prove that $\sum_{r=1}^{n} r(r+1) = \frac{n}{3}(n+1)(n+2)$.

(b) Express $\frac{1}{(x+1)(x+2)}$ in partial fractions and hence, or otherwise, prove that

$$\sum_{r=1}^{n} \frac{2}{(r+1)(r+2)} = \frac{n}{n+2}$$ [E]

33 The complex number z is given by

$$z = (1 + 3\text{i})(p + q\text{i})$$

where p and q are real and $p > 0$.

Given that $\arg z = \frac{\pi}{4}$,

(a) prove that $p + 2q = 0$.

Given also that $|z| = 10\sqrt{2}$,

(b) find the values of p and q.

(c) Write down the value of $\arg z^*$. [E]

34 (a) Sketch, for $0 < x < \frac{\pi}{2}$, the curve with equation $y = \tan x$.

By using your sketch show that the equation $\tan x = \frac{1}{x}$ has one and only one root in $0 < x < \frac{\pi}{2}$.

(b) Show further that this root lies between 0.85 and 0.87.

(c) Taking 0.85 as a first approximation to this root, use the Newton–Raphson procedure once to determine a second approximation, giving your answer to 4 decimal places. [E]

35 One root of the equation

$$z^3 - 6z^2 + 13z + k = 0$$

where k is real, is $2 + \mathrm{i}$.

Find

(a) the other roots

(b) the value of k. [E]

36 Find the complete set of values of x for which

$$\frac{1+x}{1-x} > \frac{2-x}{2+x}$$ [E]

37 Show that $\sum_{r=1}^{n} r(r+2) = \frac{n}{6}(n+1)(2n+7)$.

Using this result, or otherwise, find, in terms of n, the sum of the series

$$3\ln 2 + 4\ln 2^2 + 5\ln 2^3 + \ldots + (n+2)\ln 2^n$$

Express your answer in its simplest form. [E]

38 Given that $z = \frac{\sqrt{3}+\mathrm{i}}{1-\mathrm{i}}$, find $z^2 + \frac{1}{z^2}$, giving your answer in the form $a + \mathrm{i}b$, where a and b are real. [E]

39 Find the set of values of x for which

$$\frac{x^2+7x+10}{x+1} > 2x+7$$ [E]

40 Show that the equation $\mathrm{e}^x \cos 2x - 1 = 0$ has a root between 0.4 and 0.45.

Taking 0.45 as a first approximation to this root, apply the Newton–Raphson procedure once to obtain a second approximation, giving your answer to 3 significant figures. [E]

41 The complex number z is given by

$$z = -\sqrt{3} + \mathrm{i}$$

Find the value of

(a) $|z|$

(b) $\arg z$, giving your answer in degrees

(c) $\arg\left(\frac{\mathrm{i}}{z}\right)$, giving your answer in degrees. [E]

42 Given that $z = 2 + 3\text{i}$ is a root of the equation

$$z^3 - 6z^2 + 21z - 26 = 0$$

find the other two roots. [E]

43 Verify that the equation

$$x^3 + x - 11 = 0$$

has a root in the interval $2 \leqslant x \leqslant 3$.
By writing $x = 2 + h$, and neglecting h^2 and h^3, find an approximate value for this root, giving your answer to 2 decimal places. [E]

44 By considering $\text{f}(r) - \text{f}(r+1)$, where $\text{f}(r) = \dfrac{r+2}{r(r+1)}$, or otherwise, prove that

$$\sum_{r=1}^{n} \frac{r+4}{r(r+1)(r+2)} = \frac{3}{2} - \frac{n+3}{(n+1)(n+2)}$$ [E]

45 Sketch with the same axes the graphs of

$$y = |3x + 1| \quad \text{and} \quad y = |2 - x|$$

Find the set of values of x for which

$$|3x + 1| - |2 - x| < 3$$ [E]

46 Given that $z = 3 + 4\text{i}$, express in the form $p + q\text{i}$, where p and $q \in \mathbb{R}$,

(a) $\dfrac{1}{z}$ (b) $\dfrac{1}{z^2}$

Find the argument of $\dfrac{1}{z^2}$, giving your answer in degrees to 1 decimal place. [E]

47 Show that the equation $x^3 - 12x + 7 = 0$ has one negative real root and two positive real roots.
Show that one root α lies in the interval $[3, 3.2]$.
Use linear interpolation once on this interval to estimate α, giving your answer to 2 decimal places. Show further that your estimate is correct to 2 decimal places. [E]

48 (a) Given that $z_1 = 5 + \text{i}$ and $z_2 = -2 + 3\text{i}$,
(i) show that $|z_1|^2 = 2|z_2|^2$
(ii) find $\arg(z_1 z_2)$.
(b) Calculate, in the form $a + \text{i}b$, where $a, b \in \mathbb{R}$, the square roots of $16 - 30\text{i}$. [E]

49 Given that $f(x) \equiv 3 + 4x - x^4$, show that the equation $f(x) = 0$ has a root $x = a$, where a is in the interval $1 \leqslant a \leqslant 2$. It may be assumed that if x_n is an approximation to a, then a better approximation is given by x_{n+1}, where

$$x_{n+1} = (3 + 4x_n)^{\frac{1}{4}}$$

Starting with $x_0 = 1.75$, use this result twice to obtain the value of a to 2 decimal places. [E]

50 (a) On the same diagram, sketch the graphs of

$$y = \frac{1}{x - a} \quad \text{and} \quad y = 4|x - a|$$

where a is a positive constant. Show clearly the coordinates of any points of intersection with the coordinate axes.

(b) Hence, or otherwise, find the set of values of x for which

$$\frac{1}{x - a} < 4|x - a|$$ [E]

51 Express

$$\frac{3r + 1}{r(r - 1)(r + 1)}$$

in partial fractions. Hence, or otherwise, show that

$$\sum_{r=2}^{n} \frac{3r + 1}{r(r - 1)(r + 1)} = \frac{5}{2} - \frac{2}{n} - \frac{1}{n + 1}$$ [E]

52 Show, graphically or otherwise, that the equation $x = 2 \sin x$ has a root near $x = 1.9$. Use one iteration of the Newton–Raphson method to find a more accurate approximation to this root. [E]

53 The complex numbers z_1 and z_2 are given by

$$z_1 = 24 + 7\text{i}, \quad z_2 = 4 - 3\text{i}$$

(a) Given that $z_1 + \alpha z_2$ is real, where α is real, find the value of α.

(b) Given that $z_1 + (p + \text{i}q)z_2 = 0$, where p and q are real numbers, find p and q. [E]

54 (a) The complex numbers z_1 and z_2 are given by

$$z_1 = 57 - 17\text{i}, \quad z_2 = 5 + 6\text{i}$$

(i) Express $\frac{z_1}{z_2}$ in the form $p + q\text{i}$, where p and q are real integers.

(ii) Find $\arg z_1$, giving your answer in degrees to 1 decimal place.

(b) Given the two complex numbers w and $2w\text{i}$, state precisely

(i) the relation between their moduli

(ii) the relation between their arguments. [E]

55 The diagram shows a semicircle with O the mid-point of the diameter AB. The point P on the semicircle is such that the area of sector POB is equal to the area of the shaded segment. Angle POB is x radians.

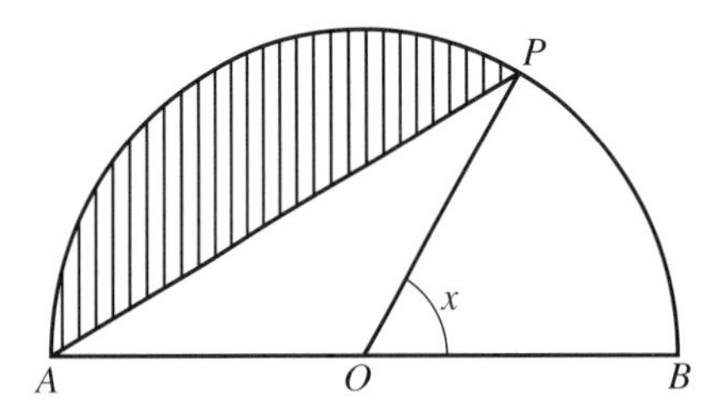

(a) Prove that $x = \frac{1}{2}(\pi - \sin x)$.

The iterative method based on the relation $x_{n+1} = \frac{1}{2}(\pi - \sin x_n)$ can be used to evaluate x.

(b) Starting with $x_1 = 1$ perform two iterations to find the values of x_2 and x_3, giving your answers to 2 decimal places. [E]

56 Given that $z_1 = 1 + 2\text{i}$ and $z_2 = \frac{3}{5} + \frac{4}{5}\text{i}$, write $z_1 z_2$ and $\frac{z_1}{z_2}$ in the form $p + \text{i}q$, where p and $q \in \mathbb{R}$. In an Argand diagram, the origin O and the points representing $z_1 z_2$, $\frac{z_1}{z_2}$, z_3 are the vertices of a rhombus. Find z_3 and sketch the rhombus on this Argand diagram.

Show that $|z_3| = \frac{6\sqrt{5}}{5}$. [E]

57 $$f(x) \equiv \ln x + x^2 - 4x$$

Show that the equation $f(x) = 0$ has a root α in the interval $[3, 4]$. Use the method of interval bisection to find α, giving your answer correct to 2 decimal places. [E]

58 A chord divides a circle, centre O, into two regions whose areas are in the ratio $2:1$. Prove that the angle θ, subtended by this chord at O, satisfies the equation $f(\theta) = 0$, where

$$f(\theta) \equiv \theta - \sin\theta - \frac{2\pi}{3}$$

Using the same axes, sketch for $0 \leqslant \theta \leqslant \pi$, the graphs of $y = \theta - \frac{2\pi}{3}$ and $y = \sin\theta$.

By taking $\frac{5\pi}{6}$ as a first approximation to the positive root of the equation $f(\theta) = 0$, apply the Newton–Raphson procedure once to obtain a second approximation, giving your answer to three decimal places. [E]

59 The complex numbers z_1 and z_2 are given by

$$z_1 = 4 + 2\text{i} \quad \text{and} \quad z_2 = -3 + \text{i}$$

(a) Display z_1 and z_2 on the same Argand diagram.

(b) Calculate $|z_1 - z_2|$.

(c) Given that $w = \frac{z_1}{z_2}$,

(i) express w in the form $a + b\text{i}$, where $a, b \in \mathbb{R}$

(ii) calculate $\arg w$, giving your answer in radians. [E]

60 In the figure, O is the centre of a circle, radius 10 cm, and the points A and B are situated on the circumference so that $\angle AOB = 2\theta$ radians. The area of the shaded segment is $44\,\text{cm}^2$. Prove that

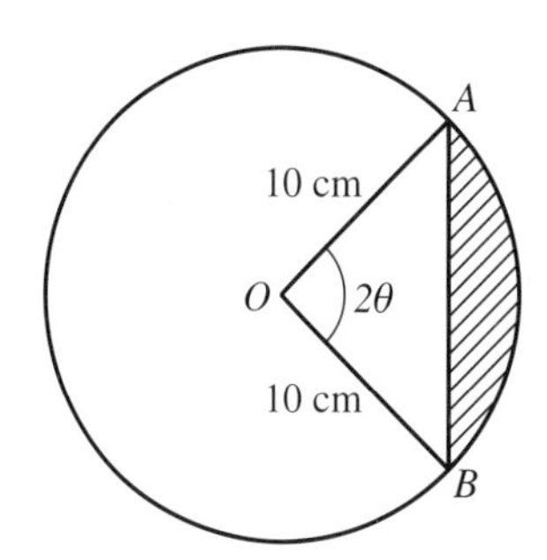

$$2\theta - \sin 2\theta - 0.88 = 0.$$

Show that a root of this equation lies between 0.9 and 1.

By taking 0.9 as a first approximation to this root, use the Newton–Raphson procedure once to determine a second approximation, giving your answer to 2 decimal places. [E]

61 (a) Using the same axes, sketch the graphs of $y = e^x$ and $y = \dfrac{1}{x}$.

(b) Deduce the number of real solutions of the equation

$$e^x = \frac{1}{x}$$

and show that this equation may be written in the form

$$x - e^{-x} = 0$$

(c) Show that the equation $x - e^{-x} = 0$ has a root in the interval $0.5 < x < 0.6$.

(d) Taking 0.6 as a first approximation to this root, use the Newton–Raphson procedure once to find a second approximation, giving your answer to 3 decimal places. [E]

62

$$f(x) \equiv 2x^3 - 5x^2 + px - 5, \quad p \in \mathbb{R}$$

The equation $f(x) = 0$ has $(1 - 2i)$ as a root. Solve the equation and determine the value of p. [E]

63 Express the complex numbers

(a) $(5 - 3i)(1 + 2i)$ (b) $\dfrac{4 - 2i}{1 - i}$

in the form $p + iq$, where p and q are real.
These numbers are represented in an Argand diagram by the points A and B respectively. Show that the point C representing the complex number $7 - i$ lies on the circle with AB as diameter. [E]

64 Find the modulus and argument of each of the complex numbers z_1 and z_2, where

$$z_1 = \frac{1 + i}{1 - i}, \quad z_2 = \frac{\sqrt{2}}{1 - i}$$

Plot the points representing z_1, z_2 and $z_1 + z_2$ on an Argand diagram. Deduce from your diagram that

$$\tan\left(\frac{3\pi}{8}\right) = 1 + \sqrt{2}$$

[E]

65 $$f(x) \equiv 2x^3 - \tan x, \quad 0 < x < \frac{\pi}{2}$$

Prove that the equation $f(x) = 0$ has two real roots and determine each of these correct to 2 decimal places.

66 (a) Given that $z = \cos\theta + i\sin\theta$, where $z \neq -1$, prove that

$$\frac{2}{1+z} = 1 - i\tan\tfrac{1}{2}\theta$$

(b) One root of the equation

$$z^3 + z^2 + 4z + \lambda = 0$$

where λ is a real number, is $1 - 3i$. Find the other roots and the value of λ. [E]

67 Given that a and b are unequal, positive numbers, prove that

$$a^3 + b^3 > a^2b + ab^2$$

First order differential equations

5

Any relation between the variables x, y, $\frac{dy}{dx}$, $\frac{d^2y}{dx^2}$, $\frac{d^3y}{dx^3}\ldots$ is called a **differential equation**. If the highest derivative that occurs in the equation is $\frac{d^ny}{dx^n}$, then the equation has **order *n***. For example, $y\frac{dy}{dx} = e^x$ is a **first order differential equation** and $\frac{d^2y}{dx^2} + 3\frac{dy}{dx} = \sin x$ is a **second order differential equation**.

5.1 The general solution of a first order differential equation in which the variables are separable

Book C3 gave the **general solution** of the first order differential equation

$$\frac{dy}{dx} = f(x)g(y)$$

as:

$$\int \frac{1}{g(y)}\,dy = \int f(x)\,dx + C$$

provided that $\frac{1}{g(y)}$ can be integrated with respect to y and that $f(x)$ can be integrated with respect to x. The constant C is called an **arbitrary constant**. Here are two examples.

Example 1

Find the general solution of the differential equation $2y^2 + xy^2 = x\frac{dy}{dx}$.

Rewrite the equation as $\frac{dy}{dx} = y^2\left(\frac{x+2}{x}\right)$,

that is:
$$\frac{1}{y^2}\frac{dy}{dx} = \left(1 + \frac{2}{x}\right)$$
$$\int \frac{1}{y^2}\,dy = \int \left(1 + \frac{2}{x}\right) dx$$
$$-\frac{1}{y} = x + 2\ln|x| + C$$

The general solution of the differential equation is
$$\frac{1}{y} = K - x - 2\ln|x|, \text{ where } K = -C$$

Example 2

Given that $y = \frac{\pi}{6}$ at $x = \frac{\pi}{6}$, solve the differential equation
$$\frac{dy}{dx} = \sin 2x \sec y$$

Since $\sec y = \frac{1}{\cos y}$ you can rewrite the equation as
$$\cos y \frac{dy}{dx} = \sin 2x$$

that is:
$$\int \cos y\,dy = \int \sin 2x\,dx$$

Integrating with respect to x, you have
$$\sin y = -\tfrac{1}{2}\cos 2x + C$$
for the general solution.

Since $y = \frac{\pi}{6}$ at $x = \frac{\pi}{6}$ this gives
$$\sin\frac{\pi}{6} = -\tfrac{1}{2}\cos\frac{\pi}{3} + C \Rightarrow \tfrac{1}{2} = -\tfrac{1}{4} + C$$

So:
$$C = \tfrac{3}{4}$$

The solution of the differential equation is
$$\sin y = \tfrac{3}{4} - \tfrac{1}{2}\cos 2x$$

Exercise 5A

Find the general solutions of the differential equations in questions 1–10.

1 $\frac{dy}{dx} = \cos 2x$

2 $\frac{dy}{dx} = \operatorname{cosec}\tfrac{1}{3}y$

3 $\tan y\frac{dy}{dx} = \cot x$

4 $\frac{dy}{dx} = e^{2y}\sec^2 x$

5 $\dfrac{dy}{dx} = \dfrac{y^2 - 1}{e^{\frac{1}{2}x}}, \quad y > 1$

6 $\dfrac{dy}{dx} = y^2 \ln x$

7 $e^{-x^2}\dfrac{dy}{dx} = xy$

8 $\dfrac{dy}{dx} = e^{x+y}$

9 $\dfrac{dy}{dx} = \dfrac{y}{x^2 - 1}, \quad x > 1$

10 $x^2\dfrac{dy}{dx} + \sin^2 y = 0$

Obtain the solution that satisfies the given conditions of the differential equations in questions **11–24**.

11 $\dfrac{dy}{dx} = 4y^2, \quad y = \frac{1}{2}$ at $x = -2$

12 $\dfrac{dy}{dx} = ye^x, \quad y = 1$ at $x = 0$

13 $\dfrac{dy}{dx} = \tan^2 x, \quad y = 0$ at $x = \frac{\pi}{4}$

14 $\dfrac{dy}{dx} = e^{2y+3x}, \quad y = \frac{1}{2}$ at $x = \frac{1}{3}$

15 $\dfrac{dy}{dx} = \dfrac{y}{x}, \quad x > 0$ and $y = 4$ at $x = 1$

16 $e^x\dfrac{dy}{dx} = y^{\frac{1}{2}}, \quad y = 4$ at $x = 0$

17 $\sin x\dfrac{dy}{dx} = e^y, \quad 0 < x < \pi, \quad y = 0$ at $x = \frac{\pi}{2}$

18 $\sin x\dfrac{dy}{dx} = \tan y(3\cos x + \sin x), \quad y = \frac{\pi}{6}$ at $x = \frac{\pi}{2}$

19 $(5 - 3\sin x)\dfrac{dy}{dx} = 40\cos x, \quad y = 0$ at $x = \frac{3\pi}{2}$

20 $(1 + \cos^2 x)\dfrac{dy}{dx} = y(y + 1)\sin 2x, \quad y = 2$ at $x = 0$

21 $y(1 + x^2)\dfrac{dy}{dx} = x(1 + y^2), \quad y = 1$ at $x = 0$

22 $\dfrac{1}{y}\dfrac{dy}{dx} = x + xy, \quad y = 1$ at $x = 0$

23 $(1 + \cos 2x)\dfrac{dy}{dx} = \sec y, \quad y = 0$ at $x = 0$

24 $e^{-x^2}\dfrac{dy}{dx} = x(y + 2)^2, \quad y = 0$ at $x = 0$

5.2 Family of solution curves

You have seen that the general solution of a first order differential equation always contains an arbitrary constant. If in addition to the differential equation a set of **boundary conditions** such as $y = 1$ at $x = 0$ is given, then you can find the arbitrary constant uniquely by substitution into the general solution. You did this in Exercise 5A, questions 11–24. Notice that different boundary conditions used in the general solution may give rise to *different* values of the arbitrary constant. So a set of specific solutions, all different from each other because the value of the arbitrary constant C is different for each, arises from one differential equation. You can sketch a graph for each of these solutions. The curves do not intersect and they are called **a family of solution curves**.

Two simple examples are used to illustrate such families of curves.

Example 3

Find the general solution of the differential equation $\frac{dy}{dx} = 2x$ and sketch the family of solution curves represented by this general solution.

Integrating $\frac{dy}{dx} = 2x$ gives:

$$y = x^2 + C$$

which is the general solution.

If $y = 0$ at $x = 0$ then $C = 0$, if $y = 1$ at $x = 0$ then $C = 1$,

if $y = 2$ at $x = 0$ then $C = 2$, if $y = -2$ at $x = 0$ then $C = -2$.

The solution curves corresponding to these different values of C look like this:

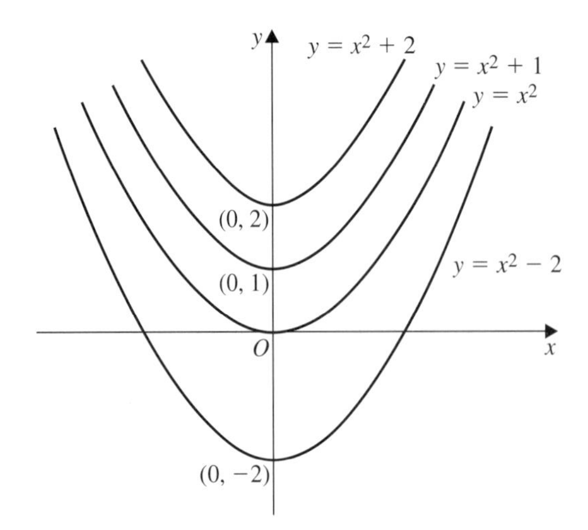

Example 4

Find the general solution of the differential equation $\frac{dy}{dx} = -\frac{x}{y}$ and interpret the solution geometrically.

By separating the variables you obtain

$$\int y\,dy = -\int x\,dx$$

That is, $\frac{1}{2}y^2 = -\frac{1}{2}x^2 + C$ is the general solution and this can be written as

$$x^2 + y^2 = a^2, \text{ where } a^2 = 2C$$

The equation $x^2 + y^2 = a^2$ represents a family of circles, all with centre at the origin and radius a, where a is a constant.

Here is the family of solution curves given by the differential equation $\frac{dy}{dx} = -\frac{x}{y}$ for $a = 1, 1\frac{1}{2}, 2$ and 3:

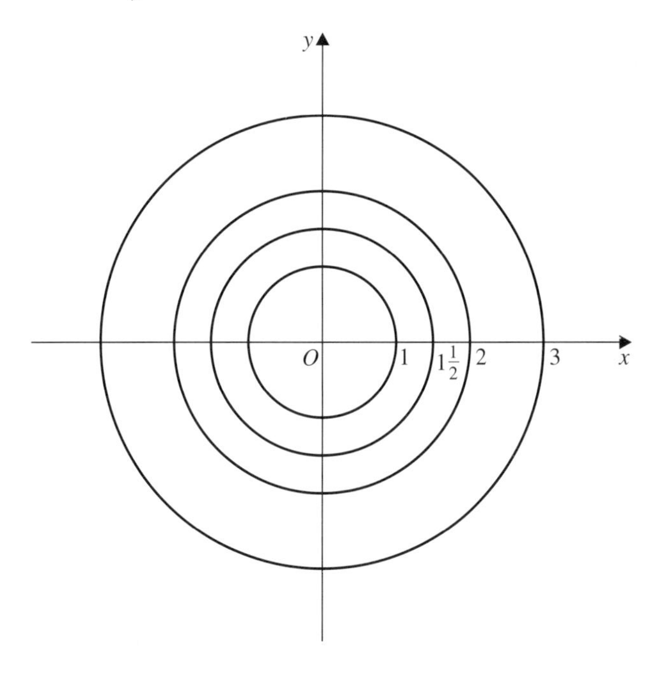

Exercise 5B

In questions **1–8**, find the general solution of the differential equation and sketch the family of solution curves represented by this general solution.

1 $\frac{dy}{dx} = 4$

2 $\frac{dy}{dx} = \frac{4}{y}$

3 $\frac{dy}{dx} = \frac{y}{x}$

4 $\frac{dy}{dx} = e^{2x}$

5 $\frac{dy}{dx} = \cos x$

6 $\frac{dy}{dx} = \frac{1-x}{y}$

7 $\frac{dy}{dx} = \frac{x}{2y}$

8 $\frac{dy}{dx} = -\frac{y}{x}, \quad x > 0$ and $y > 0$

9 Given that A is an arbitrary constant, show that $y = Ae^x$ is the general solution of the differential equation $\frac{dy}{dx} = y$. Sketch the family of solution curves for different values of A.

10 Given that $x > 0$ show that the general solution of the differential equation $\frac{dy}{dx} = \frac{y}{x \ln x}$ is $y = A \ln x$, where A is an arbitrary constant. Sketch the family of solution curves for different values of A.

5.3 First order exact differential equations

The differential equation $y + x\frac{dy}{dx} = x^3$ is first order but the variables cannot be separated. However, you should notice that the left-hand side can be written as $\frac{d}{dx}(xy)$. That, is $\frac{d}{dx}(xy) = x^3$ is another form of the equation and this form can be integrated at once to give

$$xy = \tfrac{1}{4}x^4 + C$$

as the general solution.

- **Equations of this form, where one side is the *exact* derivative of a product and the other side can be integrated with respect to x, are called *exact differential equations of the first order.***

5.4 First order linear differential equations

A **first order linear differential equation** is of the form

$$\frac{dy}{dx} + Py = Q$$

where P and Q are either functions of x or constants.

This type of first order differential equation may be exact, but more often it can be *made* exact by multiplying through the equation by a function of x. This function of x is called an **integrating factor**.

As a simple illustration consider the differential equation

$$\frac{dy}{dx} + \frac{y}{x} = x^2$$

Multiplication by x gives

$$x\frac{dy}{dx} + y = x^3 \quad \text{or} \quad y + x\frac{dy}{dx} = x^3$$

and makes the equation exact. The solution follows, as you have already seen. In this case the integrating factor is x.

Suppose now that the linear equation

$$\frac{dy}{dx} + Py = Q$$

where P and Q are functions of x, is made exact by multiplying throughout by the integrating factor $f(x)$. That is, $f(x)\frac{dy}{dx} + Pf(x)y$ is an exact derivative of a product. It follows that

$$f(x)\frac{dy}{dx} + Pf(x)y \equiv \frac{d}{dx}[yf(x)] \tag{1}$$

But

$$\frac{d}{dx}[yf(x)] = f(x)\frac{dy}{dx} + yf'(x) \tag{2}$$

by using the product rule of differentiation.

Therefore, comparing (1) and (2): $\quad Pf(x) = f'(x)$

Dividing by $f(x)$ and integrating with respect to x:

$$\int P\,dx = \int \frac{f'(x)}{f(x)}\,dx = \ln[f(x)]$$

That is, the integrating factor $f(x)$ is given by

$$f(x) = e^{\int P\,dx}$$

The linear equation $\frac{dy}{dx} + Py = Q$ can be solved by multiplying by the integrating factor $e^{\int P dx}$, provided that $e^{\int P dx}$ can be found and that the function $Qe^{\int P dx}$ can be integrated with respect to x.

So if $\frac{dy}{dx} + Py = Q$ then:

$$e^{\int P dx}\frac{dy}{dx} + Pe^{\int P dx}y = Qe^{\int P dx}$$

That is:

$$\frac{d}{dx}\left[ye^{\int P dx}\right] = Qe^{\int P dx}$$

Integrating gives the solution

$$ye^{\int P dx} = \int Qe^{\int P dx} dx + C$$

Notice that you do not have to spend a lot of time trying to integrate the left-hand side. It is always of the form ($y \times$ integrating factor).

Example 5

Find the general solution of the differential equation

$$\frac{dy}{dx} + \left(\frac{1}{x}\right)y = x^2$$

If you compare the equation with the general linear form $\frac{dy}{dx} + Py = Q$ you can see that the integrating factor is

$$e^{\int P dx} = e^{\int \frac{1}{x} dx} = e^{\ln x} = x$$

Multiplying by x gives

$$x\frac{dy}{dx} + y = x^3$$

$$\frac{d}{dx}(xy) = x^3$$

Integrating: $$xy = \tfrac{1}{4}x^4 + C$$

as you saw on page 82. The general solution is $xy = \frac{1}{4}x^4 + C$.

Example 6

Find the general solution of the differential equation

$$\cos x\frac{dy}{dx} + y\sin x = \sin x\cos^3 x$$

First write the equation in linear form by dividing by $\cos x$ to obtain

$$\frac{dy}{dx} + (\tan x)y = \sin x\cos^2 x \qquad (1)$$

The integrating factor is then

$$e^{\int \tan x \, dx} = e^{\ln \sec x} = \sec x$$

Multiply equation (1) by $\sec x$ to make it exact:

$$\sec x \frac{dy}{dx} + (\sec x \tan x)y = \sin x \cos^2 x \sec x$$

$$\frac{d}{dx}(y \sec x) = \sin x \cos x = \tfrac{1}{2}\sin 2x$$

Integrating with respect to x you obtain the general solution as

$$y \sec x = C - \tfrac{1}{4}\cos 2x$$

Example 7

Find y in terms of x given that

$$\frac{dy}{dx} - \frac{2}{x}y = x^2 \ln x, \quad x > 0$$

and that $y = 2$ at $x = 1$.

The equation $\frac{dy}{dx} - \frac{2}{x}y = x^2 \ln x$ is linear and the integrating factor is given by

$$e^{\int -\frac{2}{x} dx} = e^{-2\ln x} = e^{\ln x^{-2}} = x^{-2}$$

Multiplying the equation by x^{-2} to make it exact gives you

$$x^{-2}\frac{dy}{dx} - 2x^{-3}y = \ln x$$

That is:

$$\frac{d}{dx}(x^{-2}y) = \ln x \qquad (1)$$

Remember that $\int \ln x \, dx = x(\ln x - 1)$. This result can be obtained by integration by parts. So by integrating equation (1) you have

$$x^{-2}y = x(\ln x - 1) + C$$

which is the general solution of the differential equation.

You know also that $y = 2$ at $x = 1$ and you can find the value of C by substituting $y = 2$ and $x = 1$ in the general solution. Hence

$$2 = -1 + C \Rightarrow C = 3$$

That is:

$$x^{-2}y = x(\ln x - 1) + 3$$

Multiply by x^2 to obtain

$$y = x^3(\ln x - 1) + 3x^2$$

which is the form of the solution required.

Exercise 5C

In questions **1–5** the differential equations are exact. Find the general solution of each.

1 $y + x\frac{dy}{dx} = x^2$

2 $2xy\frac{dy}{dx} + y^2 = x^3$

3 $\frac{dy}{dx}\sin x + y\cos x = \tan x$

4 $e^{2x}\frac{dy}{dx} + 2e^{2x}y = x\sin x$

5 $\dfrac{y - x\frac{dy}{dx}}{y^2} = \cos 2x$

In questions **6–12** find the general solution of each linear differential equation.

6 $\frac{dy}{dx} + \frac{y}{x} = \cos x$

7 $\frac{dy}{dx} + \left(\frac{2}{x}\right)y = 4x + 3$

8 $\frac{dy}{dx} - \frac{y}{x} = \ln x$

9 $\frac{dy}{dx} + \frac{y}{2x} = -x^{\frac{1}{2}}$

10 $\frac{dy}{dx} + y\cot x = \cos 3x$

11 $\frac{dy}{dx} + 2y\tan x = \sin x$

12 $\frac{dy}{dx} - \frac{y}{x+1} = x$

13 Given that $x\frac{dy}{dx} - 2y = x^3 \ln x$, find y in terms of x such that $y = 2$ at $x = 1$.

14 Find y in terms of x given that

$$\frac{dy}{dx} + 2y = \sin x$$

and that the solution curve passes through the origin O.

15 Find the general solution of the differential equation

$$\frac{dy}{dx} - 2y\,\text{cosec}\,x = \tan\frac{x}{2}, \quad 0 < x < \pi$$

16 Find the general solution of the differential equation

$$\frac{dy}{dx} + 2y = e^{-2x}(x^3 + x^{-1}), \quad x > 0$$

If you also know that $y = 0$ at $x = 1$, find y in terms of x.

17 Find y in terms of x given that $x\frac{\mathrm{d}y}{\mathrm{d}x} + 3y = \mathrm{e}^x$ and that $y = 1$ at $x = 1$.

18 Solve the differential equation, giving y in terms of x, where $x^2\frac{\mathrm{d}y}{\mathrm{d}x} - xy = 1$ and $y = 2$ at $x = 1$.

SUMMARY OF KEY POINTS

1 The general solution of the differential equation $\frac{\mathrm{d}y}{\mathrm{d}x} = \mathrm{f}(x)\mathrm{g}(y)$ is

$$\int \frac{1}{\mathrm{g}(y)}\,\mathrm{d}y = \int \mathrm{f}(x)\,\mathrm{d}x + C$$

provided that $\frac{1}{\mathrm{g}(y)}$ can be integrated with respect to y and $\mathrm{f}(x)$ can be integrated with respect to x. C is an arbitrary constant.

2 In the general solution of a differential equation, different values of the arbitrary constant C, arising from different initial conditions, give rise to a series of equations whose graphs when sketched are called a family of solution curves for the differential equation.

3 An exact first order differential equation is one that can be integrated directly as it stands without any processing.

4 The first order linear differential equation

$$\frac{\mathrm{d}y}{\mathrm{d}x} + Py = Q$$

where P and Q are functions of x, is made into an exact first order differential equation by multiplying the equation by the integrating factor $\mathrm{e}^{\int P\mathrm{d}x}$, provided that $\mathrm{e}^{\int P\mathrm{d}x}$ and the integral of $\mathrm{e}^{\int P\mathrm{d}x}Q(x)$ exist.

The general solution is then

$$y\mathrm{e}^{\int P\mathrm{d}x} = \int Q\mathrm{e}^{\int P\mathrm{d}x}\,\mathrm{d}x + C$$

where C is an arbitrary constant.

Second order differential equations

6

6.1 The second order differential equation $a\frac{d^2y}{dx^2}+b\frac{dy}{dx}+cy=0$

Here you have a second order linear equation with constant coefficients a, b and c. The equation is called second order because its highest derivative of y with respect to x is $\frac{d^2y}{dx^2}$. The equation is called linear because only first degree terms in y and its derivatives occur. For example, the second order differential equation $\frac{d^2y}{dx^2}+\left(\frac{dy}{dx}\right)^3=0$ is *non*-linear because of the term $\left(\frac{dy}{dx}\right)^3$.

Suppose that $y=u$ and $y=v$ are either constants or functions of x and that they are particular distinct solutions of the differential equation

$$a\frac{d^2y}{dx^2}+b\frac{dy}{dx}+cy=0$$

where $u \neq v$. This means that u and v both satisfy the differential equation.

That is: $$a\frac{d^2u}{dx^2}+b\frac{du}{dx}+cu=0$$

and $$a\frac{d^2v}{dx^2}+b\frac{dv}{dx}+cv=0$$

When you are solving a second order differential equation, the general solution will have *two* arbitrary constants. So let's consider as a general solution:

$$y=Au+Bv,$$

where A and B are non-zero constants.

Then:

$$\frac{dy}{dx} = A\frac{du}{dx} + B\frac{dv}{dx}$$

and
$$\frac{d^2y}{dx^2} = A\frac{d^2u}{dx^2} + B\frac{d^2v}{dx^2}$$

The left-hand side of the original equation now becomes

$$a\frac{d^2y}{dx^2} + b\frac{dy}{dx} + cy$$

$$= a\left(A\frac{d^2u}{dx^2} + B\frac{d^2v}{dx^2}\right) + b\left(A\frac{du}{dx} + B\frac{dv}{dx}\right) + c(Au + Bv)$$

$$= A\left(a\frac{d^2u}{dx^2} + b\frac{du}{dx} + cu\right) + B\left(a\frac{d^2v}{dx^2} + b\frac{dv}{dx} + cv\right)$$

$$= A(0) + B(0) = 0$$

Since A and B are arbitrary constants, this proves that $y = Au + Bv$ is indeed a general solution of the differential equation.

- **The general solution of the second order differential equation**

$$\mathbf{a\frac{d^2y}{dx^2} + b\frac{dy}{dx} + cy = 0}$$

is $$\mathbf{y = Au + Bv}$$

where $y = u$ and $y = v$ are particular, distinct solutions of the differential equation.

Now you need to find the functions u and v in specific cases. This is how you do it. In the differential equation

$$a\frac{d^2y}{dx^2} + b\frac{dy}{dx} + cy = 0$$

you try as a solution $y = e^{mx}$, where m is a constant to be found.

Hence $\frac{dy}{dx} = me^{mx}$ and $\frac{d^2y}{dx^2} = m^2e^{mx}$.

If $y = e^{mx}$ is a solution of the differential equation, then

$$am^2e^{mx} + bme^{mx} + ce^{mx} = 0$$

$$\Rightarrow \quad e^{mx}(am^2 + bm + c) = 0$$

So either $e^{mx} = 0$ or $am^2 + bm + c = 0$.

But $e^{mx} > 0$ for all m, so the two values of m which you require are the roots of the quadratic equation $am^2 + bm + c = 0$. This equation in m is often called the **auxiliary quadratic equation** and it may have real, identical or complex roots. You now need to consider the different types of roots which can occur in the auxiliary quadratic equation.

The equation $am^2 + bm + c = 0$ has **real distinct roots** if $b^2 - 4ac > 0$; it has **coincident real roots** if $b^2 - 4ac = 0$; it has **complex roots** if $b^2 - 4ac < 0$. The following examples show what happens in each case.

Example 1

Find the general solution of the differential equation

$$\frac{d^2y}{dx^2} + \frac{dy}{dx} - 6y = 0$$

Take $y = e^{mx}$ as a particular solution, then $\frac{dy}{dx} = me^{mx}$ and $\frac{d^2y}{dx^2} = m^2e^{mx}$. When you substitute these into the given equation, you get:

$$m^2e^{mx} + me^{mx} - 6e^{mx} = 0$$

$$e^{mx}(m^2 + m - 6) = 0$$

$e^{mx} > 0$, so:

$$m^2 + m - 6 = 0$$

$$(m + 3)(m - 2) = 0$$

$$\Rightarrow \quad m = -3 \text{ or } m = 2$$

So two particular solutions are $y = e^{-3x}$ and $y = e^{2x}$. The general solution of the differential equation is then

$$y = Ae^{-3x} + Be^{2x}$$

where A and B constants.

- **Generalising this result, you can say that for the differential equation $a\frac{d^2y}{dx^2} + b\frac{dy}{dx} + cy = 0$, whose auxiliary quadratic equation $am^2 + bm + c = 0$ has real, distinct roots α and β, the general solution is**

$$y = Ae^{\alpha x} + Be^{\beta x}$$

where A and B are arbitrary constants.

Exercise 6A

Find the general solution of each of the following differential equations:

1 $\frac{d^2y}{dx^2} - 3\frac{dy}{dx} + 2y = 0$

2 $\frac{d^2y}{dx^2} + 4\frac{dy}{dx} + 3y = 0$

3 $\frac{d^2y}{dx^2} - 5\frac{dy}{dx} + 4y = 0$

4 $\frac{d^2y}{dx^2} + 7\frac{dy}{dx} - 18y = 0$

5 $\frac{d^2y}{dx^2} - 2\frac{dy}{dx} - 8y = 0$

6 $\frac{d^2y}{dx^2} + 5\frac{dy}{dx} + 6y = 0$

7 $3\frac{d^2y}{dx^2} - 4\frac{dy}{dx} - 4y = 0$

8 $\frac{d^2y}{dx^2} - 2\frac{dy}{dx} - 2y = 0$

9 $6\frac{d^2y}{dx^2} + 5\frac{dy}{dx} - 6y = 0$

10 $3\frac{d^2y}{dx^2} - 2\frac{dy}{dx} - 21y = 0$

Auxiliary quadratic equation with real coincident roots

Example 2

Find the general solution of the differential equation

$$\frac{d^2y}{dx^2} - 4\frac{dy}{dx} + 4y = 0$$

Take $y = e^{mx}$, then $\frac{dy}{dx} = me^{mx}$ and $\frac{d^2y}{dx^2} = m^2e^{mx}$. When you substitute these into the given equation you get:

$$m^2e^{mx} - 4me^{mx} + 4e^{mx} = 0$$
$$e^{mx}(m^2 - 4m + 4) = 0$$
$$e^{mx} > 0 \Rightarrow m^2 - 4m + 4 = 0$$
$$(m - 2)^2 = 0$$
$$m = 2 \text{ only}$$

The general solution of the differential equation in this case is $y = Ae^{2x} + Bxe^{2x}$ and you can verify this by differentiation:

$$y = Ae^{2x} + Bxe^{2x}$$

$$\frac{dy}{dx} = 2Ae^{2x} + Be^{2x} + 2Bxe^{2x}$$

$$\frac{d^2y}{dx^2} = 4Ae^{2x} + 2Be^{2x} + 2Be^{2x} + 4Bxe^{2x}$$

$$= 4Ae^{2x} + 4Be^{2x} + 4Bxe^{2x}$$

So:

$$\frac{d^2y}{dx^2} - 4\frac{dy}{dx} + 4y = 4Ae^{2x} + 4Be^{2x} + 4Bxe^{2x} - 8Ae^{2x} - 4Be^{2x} - 8Bxe^{2x} + 4Ae^{2x} + 4Bxe^{2x}$$
$$= 0$$

which shows that the differential equation $\frac{d^2y}{dx^2} - 4\frac{dy}{dx} + 4y = 0$ is satisfied by the general solution $y = (A + Bx)e^{2x}$.

- **Generalising this result, you can say that for the differential equation $a\frac{d^2y}{dx^2} + b\frac{dy}{dx} + cy = 0$, whose auxiliary quadratic equation $am^2 + bm + c = 0$ has equal roots α, the general solution is $y = (A + Bx)e^{\alpha x}$, where A and B are arbitrary constants.**

Exercise 6B

Find the general solution of each of the following differential equations:

1 $\frac{d^2y}{dx^2} - 2\frac{dy}{dx} + y = 0$

2 $\frac{d^2y}{dx^2} + 4\frac{dy}{dx} + 4y = 0$

3 $\frac{d^2y}{dx^2} - 6\frac{dy}{dx} + 9y = 0$

4 $\frac{d^2y}{dx^2} + 8\frac{dy}{dx} + 16y = 0$

5 $4\frac{d^2y}{dx^2} + 4\frac{dy}{dx} + y = 0$

6 $9\frac{d^2y}{dx^2} - 6\frac{dy}{dx} + y = 0$

7 $4\frac{d^2y}{dx^2} - 12\frac{dy}{dx} + 9y = 0$

8 $9\frac{d^2y}{dx^2} + 30\frac{dy}{dx} + 25y = 0$

9 $\frac{d^2y}{dx^2} - \sqrt{8}\frac{dy}{dx} + 2y = 0$

10 $2\frac{d^2y}{dx^2} + \sqrt{40}\frac{dy}{dx} + 5y = 0$

Auxiliary quadratic equation with pure imaginary roots

Example 3

Find the general solution of the differential equation

$$\frac{d^2y}{dx^2} + 4y = 0$$

Take $y = e^{mx}$, $\frac{dy}{dx} = me^{mx}$ and $\frac{d^2y}{dx^2} = m^2e^{mx}$ and you have on substituting into the given equation:

$$m^2e^{mx} + 4e^{mx} = 0 \Rightarrow (m^2 + 4)e^{mx} = 0$$

Since $e^{mx} > 0$ then $m^2 = -4$ or $m = \pm 2i$. The general solution is therefore

$$y = Pe^{2ix} + Qe^{-2ix}$$

where P and Q are constants.

Now it will be shown in Book P6 that

$$e^{ni\theta} = \cos n\theta + i\sin n\theta \quad \text{and} \quad e^{-ni\theta} = \cos n\theta - i\sin n\theta$$

So the general solution can be written

$$\begin{aligned} y &= P(\cos 2x + i\sin 2x) + Q(\cos 2x - i\sin 2x) \\ &= (P+Q)\cos 2x + i(P-Q)\sin 2x \end{aligned}$$

or
$$y = A\cos 2x + B\sin 2x$$

where A and B are constants and $A = P + Q$ and $B = i(P - Q)$.

- **Generalising the result, you can say that for the differential equation $\frac{d^2y}{dx^2} + n^2y = 0$, the general solution is $y = A\cos nx + B\sin nx$, where A and B are arbitrary constants.**

Auxiliary quadratic equation with complex conjugate roots

Example 4

Find the general solution of the differential equation

$$\frac{d^2y}{dx^2} - 4\frac{dy}{dx} + 13y = 0$$

Take $y = e^{mx}$, $\frac{dy}{dx} = me^{mx}$, $\frac{d^2y}{dx^2} = m^2e^{mx}$ and you have on substituting into the given equation:

$$m^2e^{mx} - 4me^{mx} + 13e^{mx} = 0$$

$$e^{mx}(m^2 - 4m + 13) = 0$$

$$e^{mx} > 0 \Rightarrow m = \frac{4 \pm \sqrt{(16-52)}}{2}$$

$$= \frac{4 \pm \sqrt{(-36)}}{2} = 2 \pm 3i$$

So the general solution is

$$y = Pe^{(2+3i)x} + Qe^{(2-3i)x}$$

where P and Q are constants,

or
$$y = e^{2x}(Pe^{3ix} + Qe^{-3ix})$$

So $y = e^{2x}(A\cos 3x + B\sin 3x)$ where $A = P + Q$ and $B = i(P - Q)$.

■ **Generalising this result, you can say that for the differential equation $a\frac{d^2y}{dx^2} + b\frac{dy}{dx} + cy = 0$, where the auxiliary quadratic equation has complex conjugate roots $p + iq$ and $p - iq$, and $p, q \in \mathbb{R}$, the general solution is**

$$y = e^{px}(A\cos qx + B\sin qx)$$

where A and B are arbitrary constants.

Exercise 6C

Find the general solution of each of the following differential equations:

1 $\frac{d^2y}{dx^2} + y = 0$

2 $\frac{d^2y}{dx^2} + 25y = 0$

3 $4\frac{d^2y}{dx^2} + 9y = 0$

4 $16\frac{d^2y}{dx^2} + 49y = 0$

5 $\frac{d^2y}{dx^2} - 2\frac{dy}{dx} + 5y = 0$

6 $\frac{d^2y}{dx^2} + 4\frac{dy}{dx} + 5y = 0$

7 $\frac{d^2y}{dx^2} - 6\frac{dy}{dx} + 10y = 0$

8 $\frac{d^2y}{dx^2} + 8\frac{dy}{dx} + 25y = 0$

9 $4\frac{d^2y}{dx^2} - 4\frac{dy}{dx} + 5y = 0$

10 $25\frac{d^2y}{dx^2} - 20\frac{dy}{dx} + 13y = 0$

6.2 The second order differential equation $a\frac{d^2y}{dx^2} + b\frac{dy}{dx} + cy = f(x)$

The method of finding a general solution to this differential equation is an extension of the work learned in section 6.1. First of all, you need to solve the differential equation

$$a\frac{d^2y}{dx^2} + b\frac{dy}{dx} + cy = 0$$

just as you did previously, and this solution is called the **complementary function**.

Next you need to find a solution of the equation $a\frac{d^2y}{dx^2} + b\frac{dy}{dx} + cy = f(x)$, where $f(x)$ could be any one of these forms:

(i) a constant, k
(ii) a linear function, $px + q$
(iii) an exponential function, ke^{px}
(iv) a trigonometric function, e.g. $p\sin x$, $q\cos 2x$ or $p\sin 3x + q\cos 3x$
(v) a quadratic function $p + qx + rx^2$.

A solution of the differential equation for any of the forms of $f(x)$ given in (i)–(v) can be found by inspection. This solution, when found, is called a **particular integral** of the equation. The following examples are typical.

Example 5

$$\frac{d^2y}{dx^2} + 3\frac{dy}{dx} + 2y = f(x)$$

Find a particular integral of this differential equation in the cases where $f(x) =$

(a) 12 (b) $3x + 5$ (c) $3e^{2x}$ (d) $\cos 2x$ (e) $5 + 12x + 2x^2$

(a) Try $y = k$ as the particular integral; then:

$$\frac{dy}{dx} = 0, \quad \frac{d^2y}{dx^2} = 0$$

and by substituting in the equation you get:

$$0 + 0 + 2k = 12 \Rightarrow k = 6$$

So $y = 6$ is the particular integral.

(b) Try $y = ax + b$ as the particular integral, then

$$\frac{dy}{dx} = a, \quad \frac{d^2y}{dx^2} = 0$$

and by substituting in the equation you get:

$$0 + 3a + 2(ax + b) \equiv 3x + 5$$

Equating x coefficients: $2a = 3 \Rightarrow a = \frac{3}{2}$

Equating constant terms: $3a + 2b = 5 \Rightarrow \frac{9}{2} + 2b = 5$

$$\Rightarrow b = \tfrac{1}{4}$$

So $y = \frac{3}{2}x + \frac{1}{4}$ is the particular integral.

(c) Try $y = ke^{2x}$ as the particular integral, then

$$\frac{dy}{dx} = 2ke^{2x} \text{ and } \frac{d^2y}{dx^2} = 4ke^{2x}$$

By substituting in the equation you get

$$4ke^{2x} + 6ke^{2x} + 2ke^{2x} \equiv 3e^{2x}$$

That is, $12k = 3 \Rightarrow k = \frac{1}{4}$

The particular integral is $y = \frac{1}{4}e^{2x}$.

(d) Try $y = a\cos 2x + b\sin 2x$, then

$$\frac{dy}{dx} = -2a\sin 2x + 2b\cos 2x$$

$$\frac{d^2y}{dx^2} = -4a\cos 2x - 4b\sin 2x$$

Substituting in the equation:

$$-4a\cos 2x - 4b\sin 2x + 3(-2a\sin 2x + 2b\cos 2x) + 2(a\cos 2x + b\sin 2x) \equiv \cos 2x$$

Equating terms in $\cos 2x$: $\quad -4a + 6b + 2a = 1 \qquad (1)$

Equating terms in $\sin 2x$: $\quad -4b - 6a + 2b = 0 \qquad (2)$

Solving equations (1) and (2) simultaneously gives

$$a = -\tfrac{1}{20} \quad \text{and} \quad b = \tfrac{3}{20}$$

The particular integral is

$$y = -\tfrac{1}{20}\cos 2x + \tfrac{3}{20}\sin 2x$$

(e) Try $y = p + qx + rx^2$, then

$$\frac{dy}{dx} = q + 2rx$$

$$\frac{d^2y}{dx^2} = 2r$$

Substituting into the equation:

$$2r + 3(q + 2rx) + 2(p + qx + rx^2) \equiv 5 + 12x + 2x^2$$

That is: $2r + 3q + 2p + (6r + 2q)x + 2rx^2 \equiv 5 + 12x + 2x^2$

Equating x^2 coefficients: $\quad 2r = 2 \Rightarrow r = 1$

Equating x coefficients: $\quad 6r + 2q = 12$

But $r = 1$, so: $\quad 6 + 2q = 12$

$$q = 3$$

Equating constant coefficients: $\quad 2r + 3q + 2p = 5$

But $r = 1$, $q = 3$ so: $\quad 2 + 9 + 2p = 5$

$$p = -3$$

The particular integral is $y = -3 + 3x + x^2$.

- **The general solution of the differential equation**

$$a\frac{d^2y}{dx^2} + b\frac{dy}{dx} + cy = f(x)$$

is: $\quad y =$ **complementary function + particular integral**

For each differential equation you need to find the complementary function and the particular integral. Then, in general, the solution is the sum of these two.

Example 6

Find the general solution of the differential equation

$$\frac{d^2y}{dx^2} + 3\frac{dy}{dx} + 2y = f(x)$$

in the cases where $f(x) =$

(a) 12 (b) $3x + 5$ (c) $3e^{2x}$ (d) $\cos 2x$ (e) $5 + 12x + 2x^2$

You can find the complementary function by solving

$$\frac{d^2y}{dx^2} + 3\frac{dy}{dx} + 2y = 0$$

Take $y = e^{mx}$, $\frac{dy}{dx} = me^{mx}$ and $\frac{d^2y}{dx^2} = m^2e^{mx}$, so by substituting in the equation you have:

$$e^{mx}(m^2 + 3m + 2) = 0$$

$e^{mx} \neq 0 \Rightarrow m = -2$ or $m = -1$

The complementary function is $Ae^{-2x} + Be^{-x}$, where A and B are arbitrary constants.

The particular integrals for parts (a), (b), (c), (d) and (e) have been found already in example 5, and you are now able to write down the general solutions as:

(a) $y = Ae^{-2x} + Be^{-x} + 6$

(b) $y = Ae^{-2x} + Be^{-x} + \frac{3}{2}x + \frac{1}{4}$

(c) $y = Ae^{-2x} + Be^{-x} + \frac{1}{4}e^{2x}$

(d) $y = Ae^{-2x} + Be^{-x} - \frac{1}{20}\cos 2x + \frac{3}{20}\sin 2x$

(e) $y = Ae^{-2x} + Be^{-x} - 3 + 3x + x^2$

Finally, in an examination question, you may be given two conditions from which the values of the arbitrary constants A and B in the general solution can be determined. Here is an example using one of the general solutions found already.

Example 7

Find y in terms of x for the differential equation

$$\frac{d^2y}{dx^2} + 3\frac{dy}{dx} + 2y = \cos 2x$$

given that $\frac{dy}{dx} = 0$ at $x = 0$ and $y = 0$ at $x = 0$.

First you find the complementary function and the particular integral, as in example 6(d), so that you can give the general solution of the differential equation as

$$y = Ae^{-2x} + Be^{-x} - \tfrac{1}{20}\cos 2x + \tfrac{3}{20}\sin 2x$$

Since $y = 0$ at $x = 0$ you have by substitution

$$0 = A + B - \tfrac{1}{20} \qquad (1)$$

Differentiating the general solution with respect to x gives

$$\frac{dy}{dx} = -2Ae^{-2x} - Be^{-x} + \tfrac{1}{10}\sin 2x + \tfrac{3}{10}\cos 2x$$

and since $\frac{dy}{dx} = 0$ at $x = 0$ you have

$$0 = -2A - B + \tfrac{3}{10} \qquad (2)$$

You can now solve equations (1) and (2) to obtain

$$A = \tfrac{1}{4}, \quad B = -\tfrac{1}{5}$$

The solution of the differential equation subject to the conditions $\frac{dy}{dx} = 0$ and $y = 0$ at $x = 0$ is

$$y = \tfrac{1}{4}e^{-2x} - \tfrac{1}{5}e^{-x} - \tfrac{1}{20}\cos 2x + \tfrac{3}{20}\sin 2x$$

Note: If the complementary function already contains a term of the type which also needs to be in the particular integral, then you must amend your trial function for the particular integral by a factor of x or x^2, as the following example illustrates.

Example 8

Find the general solution of the differential equation

$$\frac{d^2y}{dx^2} - 3\frac{dy}{dx} = 6$$

Let's deal first with the equation $\frac{d^2y}{dx^2} - 3\frac{dy}{dx} = 0$. Using $y = e^{mx}$, you have $e^{mx}(m^2 - 3m) = 0$, from which $m = 0$ or 3 and the complementary function is $A + Be^{3x}$, where A and B are constants. If you were to take the particular integral as $y = k$, then you would have the left-hand side of the equation equal to zero, which does not make sense. This occurs because the function $y = A$ is already part of the solution contained in the complementary function. So you try $y = kx$ as the possible particular integral. Then $\frac{dy}{dx} = k$ and $\frac{d^2y}{dx^2} = 0$, so you have in the equation

$$0 - 3k = 6 \Rightarrow k = -2$$

The particular integral is taken as $y = -2x$ and the general solution of the differential equation is

$$y = A + Be^{3x} - 2x$$

Example 9

Find y in terms of x given that

$$\frac{d^2y}{dx^2} - 4\frac{dy}{dx} + 4y = e^{2x}$$

and that $\frac{dy}{dx} = 1$ and $y = 0$ at $x = 0$.

Complementary function: $y = e^{mx}$, $\frac{dy}{dx} = me^{mx}$, $\frac{d^2y}{dx^2} = m^2e^{mx}$

Hence $e^{mx}(m^2 - 4m + 4) = 0 \Rightarrow m = 2$ (double root).

The complementary function is $(A + Bx)e^{2x}$.

Particular integral: As both Ae^{2x} and Bxe^{2x} are included in the complementary function, try $y = kx^2e^{2x}$ as the particular integral.

Then:
$$\frac{dy}{dx} = 2kxe^{2x} + 2kx^2e^{2x} = e^{2x}(2kx + 2kx^2)$$

$$\frac{d^2y}{dx^2} = 2ke^{2x} + 4kxe^{2x} + 4kxe^{2x} + 4kx^2e^{2x}$$
$$= e^{2x}(2k + 8kx + 4kx^2)$$

In the original equation, you have

$$\frac{d^2y}{dx^2} - 4\frac{dy}{dx} + 4y = e^{2x}$$

So: $e^{2x}(2k + 8kx + 4kx^2) - 4e^{2x}(2kx + 2kx^2) + 4e^{2x}kx^2 \equiv e^{2x}$

that is: $2ke^{2x} \equiv e^{2x}$

But $e^{2x} \neq 0 \Rightarrow 2k = 1$ and hence $k = \frac{1}{2}$.

The general solution of the differential equation is

$$y = (A + Bx)e^{2x} + \tfrac{1}{2}x^2e^{2x}$$

Since $y = 0$ at $x = 0$: $\quad 0 = (A + 0)e^0 + 0(e^0)$

$$\Rightarrow A = 0$$

So the general solution reduces to

$$y = Bxe^{2x} + \tfrac{1}{2}x^2e^{2x}$$

Differentiating $y = Bxe^{2x} + \tfrac{1}{2}x^2e^{2x}$ with respect to x:

$$\frac{dy}{dx} = Be^{2x} + 2Bxe^{2x} + xe^{2x} + x^2e^{2x}$$

and $\dfrac{dy}{dx} = 1$ at $x = 0$.

So: $\quad 1 = B + 0 + 0 + 0 \Rightarrow B = 1$

The solution subject to $\dfrac{dy}{dx} = 1$ and $y = 0$ at $x = 0$ is:

$$y = xe^{2x} + \tfrac{1}{2}x^2e^{2x}$$

Exercise 6D

Solve each of the differential equations in questions **1–15**, giving the general solution.

1 $\dfrac{d^2y}{dx^2} - 4\dfrac{dy}{dx} + 3y = 12$

2 $\dfrac{d^2y}{dx^2} + 3\dfrac{dy}{dx} + 2y = 4x$

3 $\dfrac{d^2y}{dx^2} - 2\dfrac{dy}{dx} + y = e^{2x}$

4 $\dfrac{d^2y}{dx^2} + 4\dfrac{dy}{dx} + 4y = 2x - 1$

5 $\dfrac{d^2y}{dx^2} + y = \cos 2x$

6 $\dfrac{d^2y}{dx^2} + 9y = e^{\frac{1}{2}x}$

7 $\dfrac{d^2y}{dx^2} + 4\dfrac{dy}{dx} + 5y = 10x - 12$

8 $\dfrac{d^2y}{dx^2} - 2\dfrac{dy}{dx} + 2y = \cos x$

9 $\dfrac{d^2y}{dx^2} - 4\dfrac{dy}{dx} + 3y = 6 - 3x$

10 $\dfrac{d^2y}{dx^2} + \dfrac{dy}{dx} = e^{-x}$

11 $\dfrac{d^2y}{dx^2} - 3\dfrac{dy}{dx} = 5$

12 $3\dfrac{d^2y}{dx^2} - 2\dfrac{dy}{dx} - y = x$

13 $\dfrac{d^2y}{dx^2} + 4\dfrac{dy}{dx} + 5y = \sin 2x$

14 $\dfrac{d^2y}{dx^2} + 16y = 24$

15 $4\dfrac{d^2y}{dx^2} + 4\dfrac{dy}{dx} + 2y = \sin x + \cos x$

In questions **16–25** find the solution subject to the given boundary conditions for each of the following differential equations:

16 $\dfrac{d^2y}{dx^2} - 4\dfrac{dy}{dx} + 3y = 12$; $\dfrac{dy}{dx} = 1$ and $y = 0$ at $x = 0$

17 $\dfrac{d^2y}{dx^2} + y = e^x$; $\dfrac{dy}{dx} = y = 0$ at $x = 0$

18 $\dfrac{d^2y}{dx^2} - 2\dfrac{dy}{dx} + y = \cos x$; $\dfrac{dy}{dx} = 0$ and $y = 1$ at $x = 0$

19 $\dfrac{d^2y}{dx^2} - 6\dfrac{dy}{dx} + 5y = e^{2x}$; $\dfrac{dy}{dx} = y = 2$ at $x = 0$

20 $\dfrac{d^2y}{dx^2} + 2\dfrac{dy}{dx} + 2y = 4x$; $\dfrac{dy}{dx} = y = 0$ at $x = 0$

21 $\dfrac{d^2y}{dx^2} + 4\dfrac{dy}{dx} + 4y = 2x + 4$; $y = 1$, $\dfrac{dy}{dx} = 0$ at $x = 0$

22 $\dfrac{d^2y}{dx^2} + 2\dfrac{dy}{dx} + 10y = 20x - 6$; $y = 0$, $\dfrac{dy}{dx} = 6$ at $x = 0$

23 $\dfrac{d^2y}{dx^2} + 6\dfrac{dy}{dx} + 25y = 6\sin x$; $\dfrac{dy}{dx} = y = 0$ at $x = 0$

24 $\dfrac{d^2y}{dx^2} + 9y = 8\sin x$; $\dfrac{dy}{dx} = y = 0$ at $x = \dfrac{\pi}{2}$

25 $\dfrac{d^2y}{dx^2} - 7\dfrac{dy}{dx} + 6y = 36x$; $\dfrac{dy}{dx} = 4$ and $y = 0$ at $x = 0$

26 Show that $\frac{1}{2}x\sin x$ is a particular integral of the differential equation

$$\frac{d^2y}{dx^2} + y = \cos x$$

Hence find the general solution.

27 Find the value of the constant k so that kxe^{2x} is a particular integral of the differential equation

$$\frac{d^2y}{dx^2} - 14\frac{dy}{dx} + 24y = 4e^{2x}$$

Hence find y in terms of x, given that $y = 0$ and $\dfrac{dy}{dx} = 0$ at $x = 0$.

28 Find y in terms of x given that

$$\frac{d^2y}{dx^2} + 2\frac{dy}{dx} + 5y = 20e^{-x}$$

and that $\frac{dy}{dx} = 3$ and $y = 1$ at $x = 0$.

29 Find the general solution of the differential equation

$$4\frac{d^2y}{dx^2} - 5\frac{dy}{dx} + y = 17(\cos x - \sin x)$$

30 For the differential equation $\frac{d^2y}{dx^2} + 4y = 10e^{-x}$ find the solution for which $\frac{dy}{dx} = -1$ and $y = \frac{1}{2}$ at $x = 0$.

6.3 Solving differential equations using a change of variable

Some first order and second order differential equations can be transformed by a substitution that changes one of the variables in the equation into a new variable. You can often use substitutions to move from an equation that cannot be easily integrated to a transformed equation which you know how to solve from work already covered in this chapter. Examination questions give you the substitution needed. The following examples are typical.

Example 10

Use the substitution $y = zx$, where z is a function of x, to transform the differential equation

$$x^2\frac{dy}{dx} = y(x + y)$$

into a differential equation in z and x. By first solving this equation, find y in terms of x for $x > 0$, given that $y = -1$ at $x = 1$.

Since $y = zx$, the product rule of differentiation gives you

$$\frac{dy}{dx} = z + x\frac{dz}{dx}$$

Substituting into the equation:

$$x^2\frac{dy}{dx} = y(x + y)$$

gives: $$x^2\left(z + x\frac{\mathrm{d}z}{\mathrm{d}x}\right) = zx(x + zx)$$

$$x^2z + x^3\frac{\mathrm{d}z}{\mathrm{d}x} = zx^2 + z^2x^2$$

or: $$x^3\frac{\mathrm{d}z}{\mathrm{d}x} = z^2x^2$$

That is: $$\frac{\mathrm{d}z}{\mathrm{d}x} = \frac{z^2}{x}$$

since $x > 0$.

This is a first order differential equation in which the variables can be separated to give

$$\int \frac{1}{z^2}\,\mathrm{d}z = \int \frac{1}{x}\,\mathrm{d}x$$

So: $$-\frac{1}{z} = \ln x + C$$

where the modulus sign has been omitted since $x > 0$.

Now $y = zx \Rightarrow z = \frac{y}{x}$

and hence: $$\frac{1}{z} = \frac{x}{y}$$

So: $$-\frac{x}{y} = \ln x + C$$

But $y = -1$ at $x = 1$.

Thus: $$1 - \ln 1 = C \Rightarrow C = 1$$

So the equation becomes

$$-\frac{x}{y} = \ln x + 1$$

$$-\frac{y}{x} = \frac{1}{1 + \ln x}$$

That is $y = \dfrac{-x}{1 + \ln x}$ is the particular solution required.

Example 11

Given that $x = \mathrm{e}^u$, where u is a function of x, show that

(a) $x\dfrac{\mathrm{d}y}{\mathrm{d}x} = \dfrac{\mathrm{d}y}{\mathrm{d}u}$ (b) $x^2\dfrac{\mathrm{d}^2y}{\mathrm{d}x^2} = \dfrac{\mathrm{d}^2y}{\mathrm{d}u^2} - \dfrac{\mathrm{d}y}{\mathrm{d}u}$

Hence find the general solution of the differential equation

$$x^2\frac{\mathrm{d}^2y}{\mathrm{d}x^2} - 5x\frac{\mathrm{d}y}{\mathrm{d}x} + 9y = 0$$

(a) $x = e^u \Rightarrow \dfrac{dx}{du} = e^u$

Using the chain rule:

$$\frac{dy}{du} = \frac{dy}{dx} \cdot \frac{dx}{du} = e^u \frac{dy}{dx} = x\frac{dy}{dx}, \text{ as required.}$$

(b)
$$\begin{aligned}\frac{d^2y}{du^2} &= \frac{d}{du}\left(\frac{dy}{du}\right)\\ &= \frac{d}{du}\left(e^u \frac{dy}{dx}\right)\\ &= e^u \frac{dy}{dx} + e^u \frac{d^2y}{dx^2} \cdot \frac{dx}{du}\\ &= \frac{dy}{du} + x^2\frac{d^2y}{dx^2}, \text{ since } \frac{dx}{du} = e^u \text{ and } e^u = x\end{aligned}$$

So: $x^2\dfrac{d^2y}{dx^2} = \dfrac{d^2y}{du^2} - \dfrac{dy}{du}$, as required.

Put these results from (a) and (b) into the differential equation

$$x^2\frac{d^2y}{dx^2} - 5x\frac{dy}{dx} + 9y = 0$$

to obtain:
$$\frac{d^2y}{du^2} - \frac{dy}{du} - 5\frac{dy}{du} + 9y = 0$$

which is a second order differential equation with constant coefficients: you know how to solve it.

Put $y = e^{mu}$, $\dfrac{dy}{du} = me^{mu}$ and $\dfrac{d^2y}{du^2} = m^2e^{mu}$. Substituting these gives the auxiliary quadratic equation as

$$m^2 - 6m + 9 = 0$$

$$\Rightarrow \text{double root } m = 3$$

The general solution of the differential equation

$$\frac{d^2y}{du^2} - 6\frac{dy}{du} + 9y = 0$$

is
$$y = (A + Bu)e^{3u}$$

where A and B are arbitrary constants.

$x = e^u \Rightarrow u = \ln x$ and the general solution of the differential equation

$$x^2\frac{d^2y}{dx^2} - 5x\frac{dy}{dx} + 9y = 0$$

is
$$y = (A + B\ln x)e^{3\ln x}$$

that is:
$$y = (A + B\ln x)x^3$$

Exercise 6E

In questions **1–7** find the general solution of each differential equation using the substitution given, where v is a function of x.

1 $\dfrac{dy}{dx} = \dfrac{x}{y} + \dfrac{y}{x}$, $x > 0$, $y > 0$; $y = vx$

2 $x\dfrac{dy}{dx} = x + y$, $x > 0$; $y = vx$

3 $\dfrac{dy}{dx} = \dfrac{y(x + 2y)}{x(y + 2x)}$, $x \neq 0$; $y = vx$

4 $(x + y)\dfrac{dy}{dx} = x^2 + xy + x + 1$; $y = v - x$

5 $x^2\dfrac{d^2y}{dx^2} + x\dfrac{dy}{dx} + y = 0$; $x = e^v$

6 $x^2\dfrac{d^2y}{dx^2} - 4x\dfrac{dy}{dx} + 6y = 0$; $x = e^v$

7 $x\dfrac{d^2y}{dx^2} - 2\dfrac{dy}{dx} + x = 0$; $\dfrac{dy}{dx} = v$

8 Given that $y = 1$ at $x = 2$, use the substitution $v = 3x - y - 3$ to solve the differential equation

$$(3x - y - 1)\frac{dy}{dx} = (3x - y + 3)$$

9 Use the substitution $v = y^{-2}$ to find the general solution of the differential equation $\dfrac{dy}{dx} + \dfrac{y}{x} = x^2y^3$

Find also y in terms of x, given that $y = 1$ at $x = 1$.

SUMMARY OF KEY POINTS

1 For the second order differential equation

$$a\frac{d^2y}{dx^2} + b\frac{dy}{dx} + cy = 0$$

the auxiliary quadratic equation is

$$am^2 + bm + c = 0$$

(i) If the auxiliary quadratic equation has real distinct roots α and β (condition $b^2 > 4ac$), then the general solution is

$$y = Ae^{\alpha x} + Be^{\beta x}$$

where A and B are constants.

(ii) If the auxiliary quadratic equation has real coincident roots α (condition $b^2 = 4ac$), then the general solution is

$$y = (A + Bx)e^{\alpha x}$$

where A and B are constants.

(iii) If the auxiliary quadratic equation has pure imaginary roots $\pm ni$, arising from $m^2 + n^2 = 0$, the general solution is

$$y = A\cos nx + B\sin nx$$

where A and B are constants and $n \in \mathbb{R}$.

(iv) If the auxiliary quadratic equation has complex conjugate roots $p \pm iq$, $p, q \in \mathbb{R}$ (condition $b^2 < 4ac$), the general solution is

$$y = e^{px}[A\cos qx + B\sin qx]$$

where A and B are constants.

2 For the differential equation

$$a\frac{d^2y}{dx^2} + b\frac{dy}{dx} + cy = f(x)$$

where a, b and c are constants, the *complementary function* is the general solution of the differential equation $a\frac{d^2y}{dx^2} + b\frac{dy}{dx} + cy = 0$ and a *particular integral* is any solution (i.e. function of x) that satisfies the differential equation

$$a\frac{d^2y}{dx^2} + b\frac{dy}{dx} + cy = f(x)$$

The general solution of the differential equation is

$y =$ complementary function + particular integral

3 A change of variable given by a substitution can transform a differential equation from one in say (x, y) which is not immediately integrable into a differential equation in say (x, v) which *is* immediately integrable or a recognised equation for which a method of solution has already been learned.

Polar coordinates

7

7.1 Polar coordinates

Cartesian coordinates are the most used system for fixing the position of points in a plane but some curves have very complicated cartesian equations and in such cases a different coordinate system may be more useful. **Polar coordinates** are an alternative system which you will find useful when studying certain curves.

In the diagram, the position of the point P is given by the cartesian coordinates (x, y) where $OQ = x$ and $PQ = y$ with the usual sign conventions for x and y as P takes up positions in the other quadrants as defined by the coordinate axes Ox and Oy. In a cartesian coordinate system the point O is, of course, called the origin.

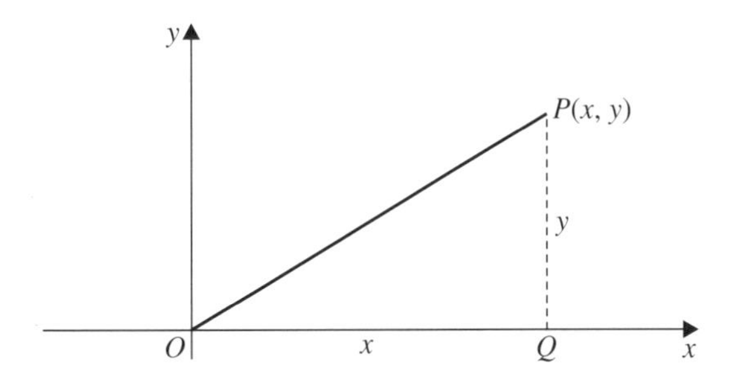

If, however, you take OP to be of length r and the angle POQ to be θ radians, then provided you restrict r to be positive and you restrict θ to the interval $-\pi < \theta \leqslant \pi$, then (r, θ) are called the **polar coordinates** of the point P. The point O is called the **pole** and, in this case, the line Ox is called the **initial line**. The initial line is denoted by the letter l, and is taken to be the positive x-axis.

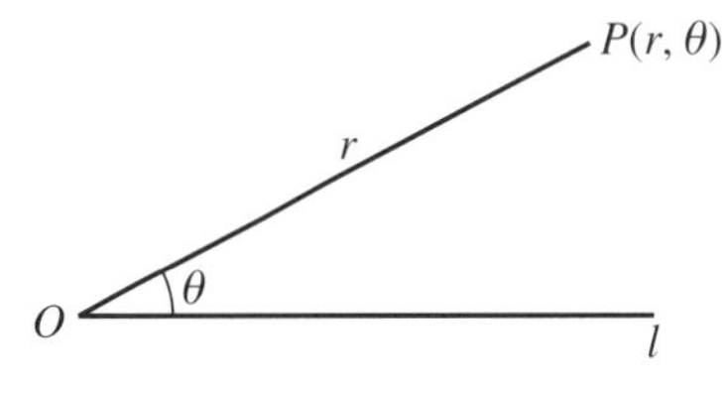

In polar coordinate systems, the angle θ is always measured from the initial line and the usual sign convention prevails: that is, a counter-clockwise rotation from l is positive.

The points $A(2, \frac{2\pi}{3})$ and $B(3, -\frac{3\pi}{4})$ are shown:

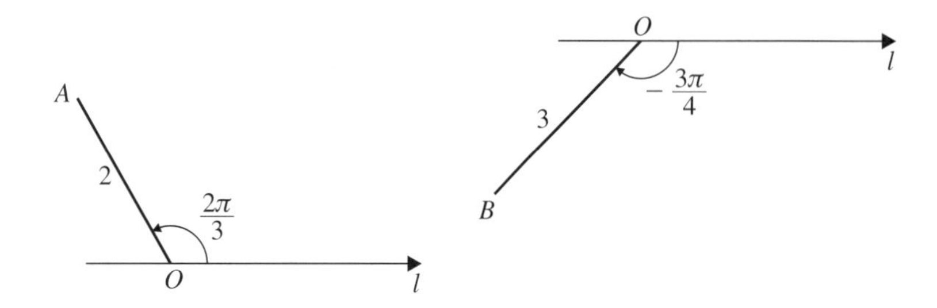

Provided that you make the origin and pole coincide and take the initial line to be the positive x-axis, it is easy to find relations between the cartesian coordinates (x, y) of a point P and the polar coordinates (r, θ) of P.

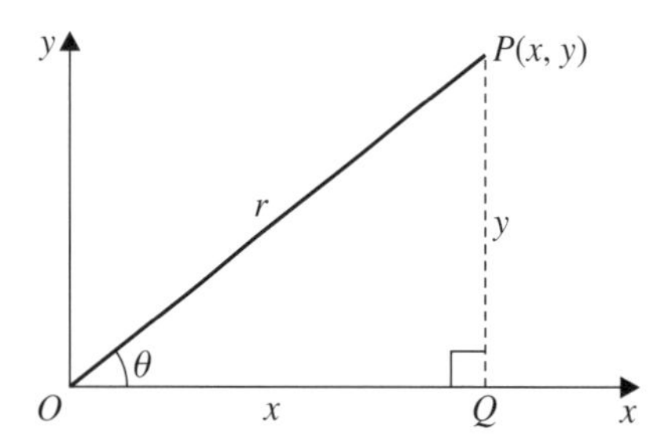

Using basic trigonometry in $\triangle OPQ$, you have

$$x = r\cos\theta \quad \text{and} \quad y = r\sin\theta$$

Also
$$r = (x^2 + y^2)^{\frac{1}{2}} \quad \text{and} \quad \theta = \arctan\left(\frac{y}{x}\right)$$

where the sign of y and the sign of x determine in which quadrant θ lies. Remember the restrictions that r is positive (or possibly zero) and that $-\pi < \theta \leqslant \pi$. (This interval may also be represented as $(-\pi, \pi]$ – see the list of symbols and notation at the end of the book.)

An alternative convention is to restrict θ to the interval $0 \leqslant \theta < 2\pi$, which can be written as $[0, 2\pi)$

With the restrictions in place you have *both* x and y unique in terms of r and θ *and* r and θ unique in terms of x and y.

You should, however, be aware that other textbooks and syllabuses may allow negative values of r to be used. For example, one convention states that negative values of r are defined as 'measured from the pole opposite to the direction of r positive'. Then the coordinates $(2, \frac{\pi}{3})$ and $(-2, -\frac{2\pi}{3})$ would represent the same point and the uniqueness of r and θ for given (x, y) is lost.

7.2 Curve sketching in polar coordinates

Normally you will be asked to *sketch* rather than accurately plot a curve given in polar coordinates, but you can easily make up polar coordinate graph paper like this if you need it.

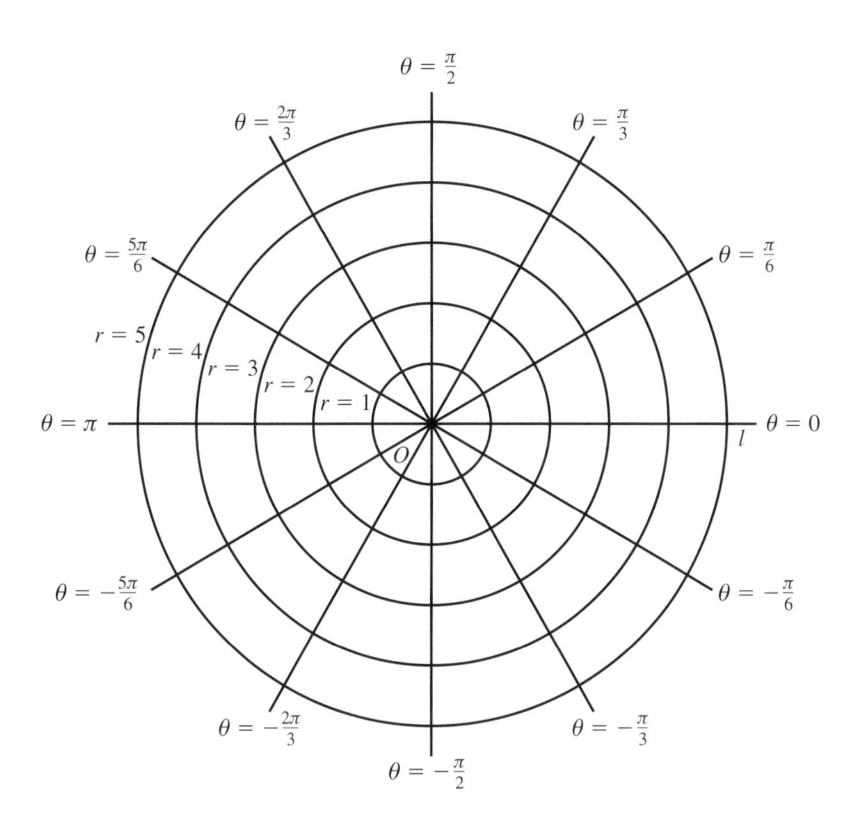

Often the general shape of a curve with equation $f(r, \theta) = 0$ is best investigated by finding the polar coordinates of a number of points on the curve. Here are some simple illustrations of how to form the polar equation of a straight line and a circle.

(i) The equation $r = c$, where c is a positive constant, represents a *circle* centre O, radius c.

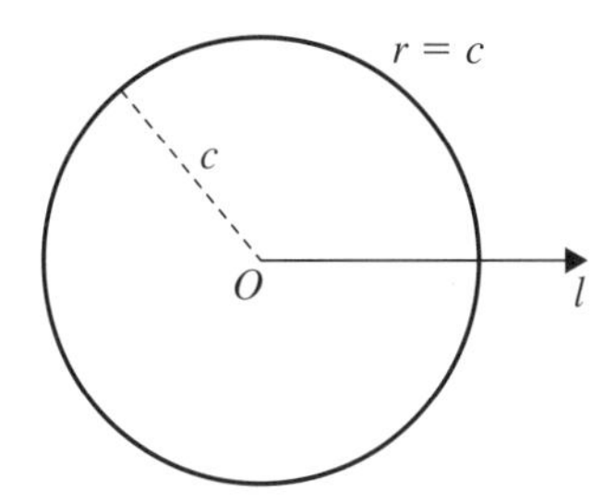

(ii) The equation $\theta = \alpha$, where α is a constant, represents the *half-line* from O at angle α radians to the initial line.

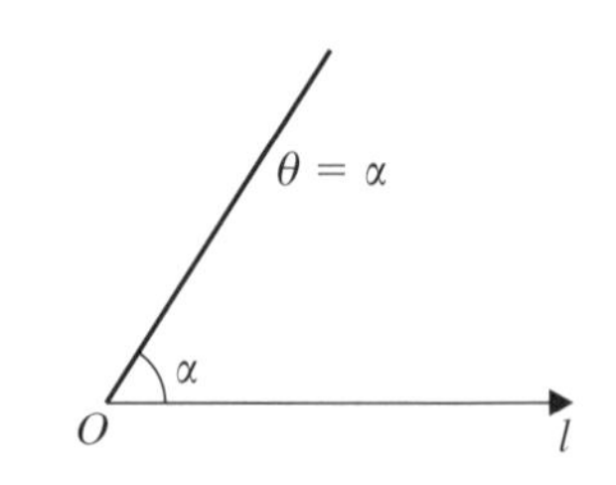

(iii) Circle centre $(b, 0)$ (in cartesian coordinates), radius a. By the cosine rule, for any point $P(r, \theta)$ on the circle you have

$$a^2 = r^2 + b^2 - 2br\cos\theta$$

which is the polar equation of the circle.

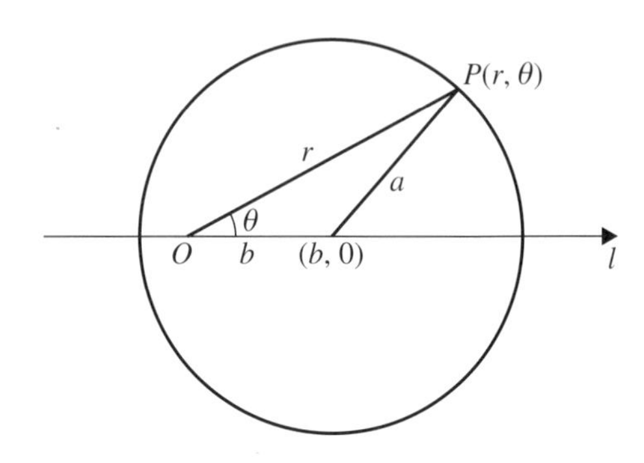

(iv) Straight line at a perpendicular distance p from the pole O, where the perpendicular OA makes a fixed angle α with l.

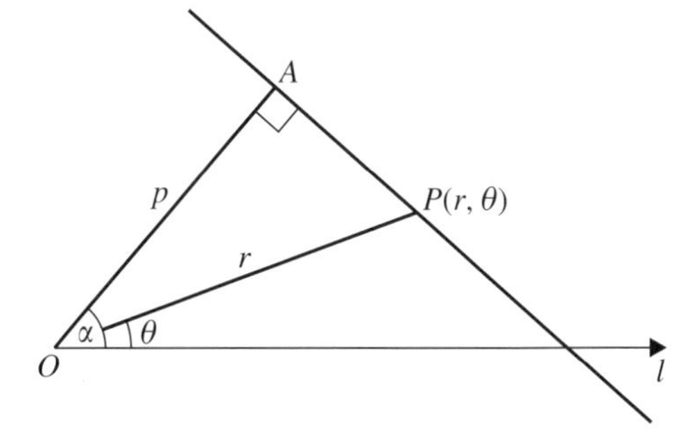

For any point $P(r, \theta)$ on the line you have

$$\cos \angle POA = \frac{OA}{OP}$$

That is: $$\cos(\alpha - \theta) = \frac{p}{r}$$

So $p = r\cos(\alpha - \theta)$ is the polar equation of the line AP.

Example 1

Sketch the curve with polar equation $r = 2a(1 + \cos\theta)$, where a is a positive constant.

It is worth remarking at once that you can see by inspection that $f(\theta) \equiv 2a(1 + \cos\theta) \equiv f(-\theta)$, because f is an even function (since cos is an even function). This means that the graph is symmetrical about the initial line, so you only need to consider values of θ from 0 to π, some of which are shown with corresponding values of r in the table:

θ	0	$\frac{\pi}{6}$	$\frac{\pi}{3}$	$\frac{\pi}{2}$	$\frac{2\pi}{3}$	$\frac{5\pi}{6}$	π
r	$4a$	$3.73a$	$3a$	$2a$	a	$0.268a$	0

If you plot these values and those below the initial line you get the curve shown, which is called a **cardioid**.

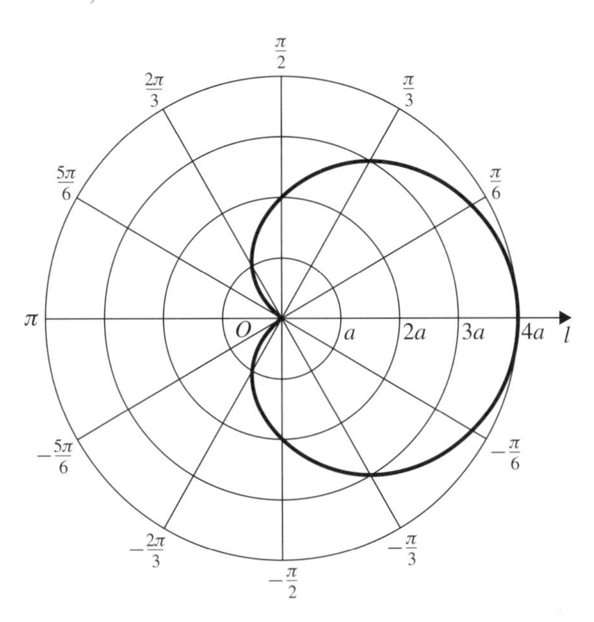

Example 2

Sketch the curve with polar equation $r = 2\sin 3\theta$.

Since $\sin 3\theta \leqslant 1$ it follows that $r \leqslant 2$ and that $r = 2$ at $\sin 3\theta = 1$; that is, where $\theta = \frac{\pi}{6}, \frac{5\pi}{6}$ and $-\frac{\pi}{2}$. So the curve touches the circle with equation $r = 2$ at $(2, \frac{\pi}{6})$, $(2, \frac{5\pi}{6})$ and $(2, -\frac{\pi}{2})$ and otherwise lies inside this circle.

The curve does not exist when r is negative, that is when $\sin 3\theta < 0$, since r is defined as being positive or zero.

$\sin 3\theta < 0$ for the intervals $\frac{\pi}{3} < \theta < \frac{2\pi}{3}$, $-\pi < \theta < -\frac{2\pi}{3}$ and $-\frac{\pi}{3} < \theta < 0$.

For $0 \leqslant \theta \leqslant \frac{\pi}{6}$, r increases from 0 to 2.

For $\frac{\pi}{6} \leqslant \theta \leqslant \frac{\pi}{3}$, r decreases from 2 to 0.

This pattern repeats itself similarly in the other two intervals where the curve exists, as shown in the sketch. The circle $r = 2$ is shown by the dotted curve. The curve with equation $r = 2\sin 3\theta$ has 3 'leaves' or 'petals'.

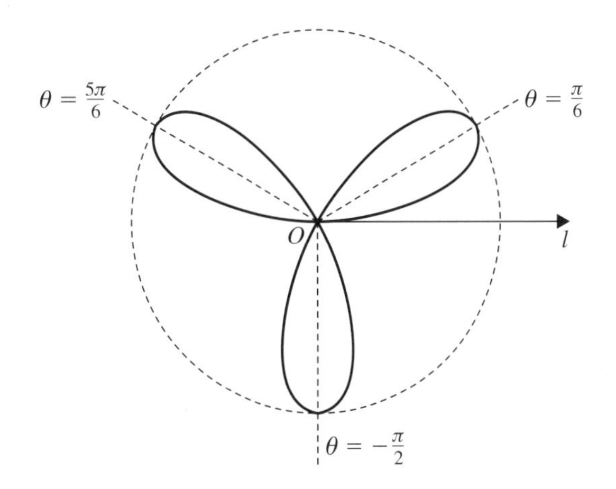

7.3 Transforming from a polar to a cartesian equation, and conversely

If you take the pole as origin and the initial line as the positive x-axis, you have for the point $P(r, \theta)$ in polar coordinates and $P(x, y)$ in cartesian coordinates

$$r\cos\theta = x \quad \text{and} \quad r\sin\theta = y$$

Example 3
Transform the polar equation $r^2 = 4\cos 2\theta$ into its cartesian form.

Since $\cos 2\theta \equiv \cos^2\theta - \sin^2\theta$, $r^2 = x^2 + y^2$, $\cos\theta = \dfrac{x}{r}$ and $\sin\theta = \dfrac{y}{r}$, you have for the equation $r^2 = 4\cos 2\theta$:

$$x^2 + y^2 = 4\left(\frac{x^2}{r^2} - \frac{y^2}{r^2}\right)$$

That is:
$$r^2(x^2 + y^2) = 4(x^2 - y^2)$$

and replacing r^2 by $x^2 + y^2$ again you get

$$(x^2 + y^2)^2 = 4(x^2 - y^2)$$

as the cartesian equation.

Example 4
Transform the cartesian equation

$$x^2 + 3y^2 = 3$$

into a polar equation of the form

$$r^2 = \frac{a}{b + c\cos 2\theta}$$

identifying the integers a, b and c.

Take $x = r\cos\theta$ and $y = r\sin\theta$ in the cartesian equation to obtain the polar equation:

$$r^2\cos^2\theta + 3r^2\sin^2\theta = 3$$

From the identities $\cos 2\theta \equiv 2\cos^2\theta - 1 \equiv 1 - 2\sin^2\theta$ you have

$$\cos^2\theta \equiv \tfrac{1}{2}(1 + \cos 2\theta), \quad \sin^2\theta \equiv \tfrac{1}{2}(1 - \cos 2\theta)$$

The polar equation reduces to

$$\tfrac{1}{2}r^2(1 + \cos 2\theta) + \tfrac{3}{2}r^2(1 - \cos 2\theta) = 3$$

That is:
$$r^2(2 - \cos 2\theta) = 3$$

which is $r^2 = \dfrac{3}{2 - \cos 2\theta}$ in the form required.

Note: Sketches of the curves discussed in examples 3 and 4 are shown below for you to refer to. The curves are called a **lemniscate** and an **ellipse** respectively.

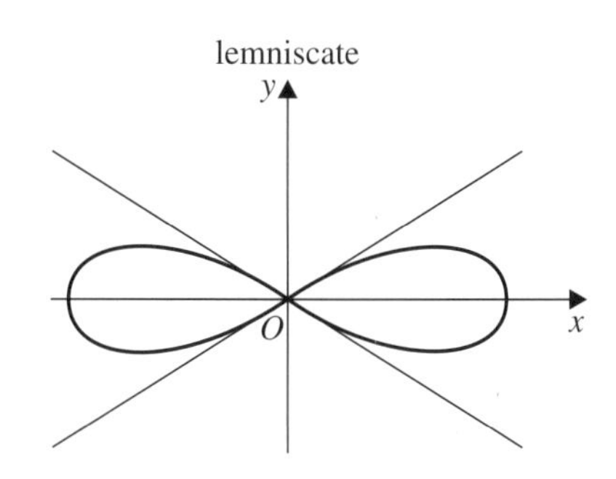

$r^2 = 4 \cos 2\theta$

$(x^2 + y^2)^2 = 4(x^2 - y^2)$

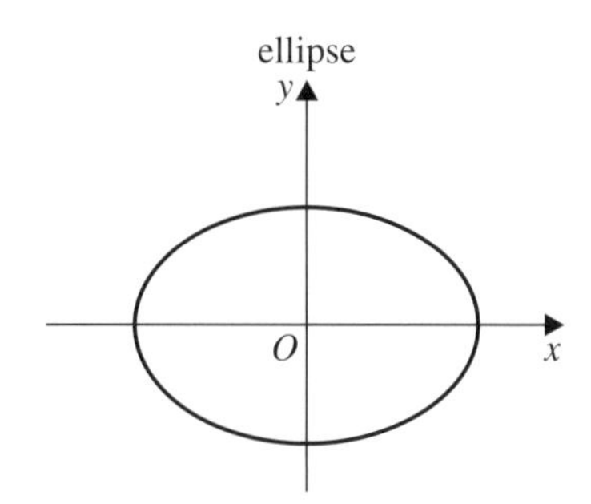

$x^2 + 3y^2 = 3$

$r^2 = \dfrac{3}{2 - \cos 2\theta}$

Exercise 7A

1 Given that the origin O and the pole coincide and that the initial line is Ox, find the polar coordinates of the points whose cartesian coordinates are:

(a) $(3, 0)$ (b) $(0, 4)$ (c) $(-3.5, 0)$
(d) $(0, -5)$ (e) $(3, 4)$ (f) $(-4, 3)$
(g) $(-1, -1)$ (h) $(5, -12)$ (i) $(-1.2, -0.9)$
(j) $(\sin\frac{\pi}{3}, \cos\frac{\pi}{3})$

2 Given that the pole O and the origin coincide and that the positive x-axis is the initial line, find the cartesian coordinates of the points whose polar coordinates are:

(a) $(3, 0)$ (b) $(2, \frac{\pi}{2})$ (c) $(5, -\frac{\pi}{2})$
(d) $(2, \pi)$ (e) $(3, \frac{\pi}{3})$ (f) $(4, \frac{2\pi}{3})$
(g) $(4, -\frac{\pi}{6})$ (h) $(4, -\frac{5\pi}{6})$ (i) $(3, -1)$
(j) $(2, -2)$

3 Sketch the lines (or half-lines) with polar equations:

(a) $\theta = -\frac{\pi}{3}$ (b) $\theta = 3$ (c) $r\cos\theta = 2$
(d) $r\sin\theta = -2$ (e) $r\cos(\theta - \frac{\pi}{6}) = 2$ (f) $r\cos(\theta + \frac{\pi}{6}) = 4$

4 Sketch the following, where a is a positive constant:

(a) the parabola with equation $r = \dfrac{a}{1 + \cos\theta}$, $-\frac{\pi}{2} < \theta < \frac{\pi}{2}$

(b) the ellipse with equation $r = \dfrac{2a}{2 + \cos\theta}$, $\quad -\pi < \theta \leqslant \pi$

(c) the spiral with equation $r = a\theta$, $\quad 0 \leqslant \theta < 2\pi$.

5 Sketch the cardioid with polar equation $r = 2(1 + \cos\theta)$. Prove that for all the chords POQ of the cardioid which pass through the pole O, the length of PQ is constant.

6 The polar equation of a curve C is $r^2 = 2\sin 2\theta$.

(a) Find a cartesian equation for C.

(b) Show that C could be represented parametrically by the equations

$$x = \frac{2t}{1+t^4}, \quad y = \frac{2t^3}{1+t^4}$$

where t is a parameter.

7 Transform to polar form the circles with equations given in cartesian form by

$$x^2 + y^2 + 2x = 0 \quad \text{and} \quad x^2 + y^2 - 2y = 0$$

8 Obtain the polar form of the following cartesian equations where a is a positive constant.

(a) $y^2 = x(a - x)$ (b) $xy = 4a^2$ (c) $(x^2 + y^2)^2 = 2a^2xy$

9 Obtain the cartesian form of the following polar equations, where a is a positive constant.

(a) $r = a\tan\theta\sec\theta$ (b) $r = 2a\tan\theta$ (c) $r = a(1 - \cos\theta)$

10 Given that a is a positive constant, sketch the curve with polar equation $r^2 = 4a^2\cos 3\theta$.

7.4 Areas of regions expressed in polar coordinates

In the diagram, the points A and B, referred to O as pole and l as initial line, have polar coordinates (r_A, α) and (r_B, β) respectively on the curve with equation $f(r, \theta) = 0$. You want to find the area of the region bounded by the half-lines $\theta = \alpha$ and $\theta = \beta$ and the arc of the curve $f(r, \theta) = 0$ between A and B. Assume that r increases with θ throughout the interval from A to B.

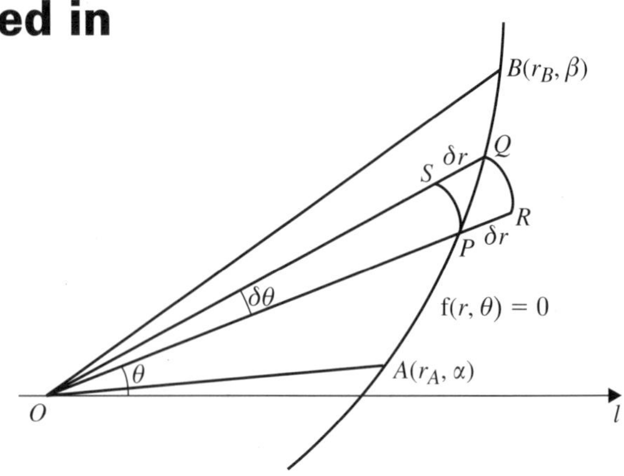

Take a point $P(r, \theta)$ and a neighbouring point $Q(r + \delta r, \theta + \delta\theta)$ on the curve. The circle with centre O and radius r cuts OQ at the point S. The circle with centre O and radius $r + \delta r$ cuts OP produced at R, as shown in the diagram. Now remember that the area of a circular sector, radius a and angle ϕ radians, is $\frac{1}{2}a^2\phi$ (Book C1, page 85). So for the required region OPQ:

$$\text{area of sector } OPS < \text{area } OPQ < \text{area of sector } ORQ$$

$$\Rightarrow \quad \tfrac{1}{2}r^2\delta\theta < \text{area } OPQ < \tfrac{1}{2}(r + \delta r)^2\delta\theta$$

Therefore, the area of the required region OAB is given by

$$\sum_{\theta=\alpha}^{\theta=\beta} \tfrac{1}{2}r^2\delta\theta < \text{area } OAB < \sum_{\theta=\alpha}^{\theta=\beta} \tfrac{1}{2}(r + \delta r)^2\,\delta\theta$$

As $\delta\theta \to 0$, $\delta r \to 0$,

and so $$\sum_{\theta=\alpha}^{\theta=\beta} \tfrac{1}{2}r^2\delta\theta \to \int_{\alpha}^{\beta} \tfrac{1}{2}r^2\,\mathrm{d}\theta$$

and $$\sum_{\theta=\alpha}^{\theta=\beta} \tfrac{1}{2}(r + \delta r)^2\,\delta\theta \to \int_{\alpha}^{\beta} \tfrac{1}{2}r^2\,\mathrm{d}\theta$$

■ So: $$\textbf{area of region } \boldsymbol{OAB} = \int_{\alpha}^{\beta} \tfrac{1}{2}\boldsymbol{r}^2\,\mathbf{d}\boldsymbol{\theta}$$

You should memorise this important formula.

Example 5

Find the area of the cardioid with polar equation $r = 2a(1 + \cos\theta)$, the sketch of which was shown in example 1.

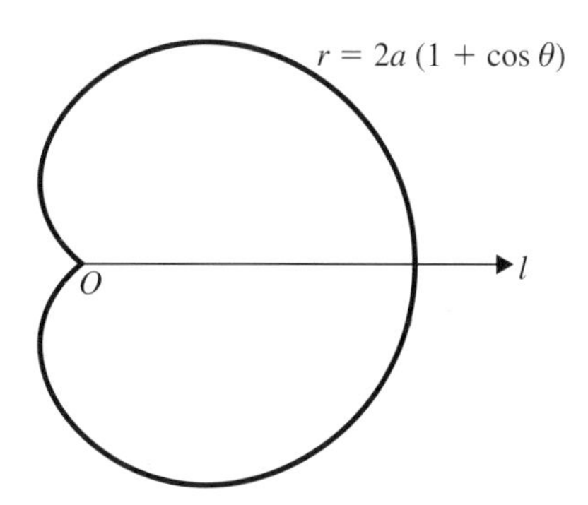

As the curve is symmetrical about the initial line, the area of the cardioid is twice the area of the part above the initial line.

So:

$$\text{area of cardioid} = 2\int_0^{\pi} \tfrac{1}{2}r^2 \, d\theta$$
$$= \int_0^{\pi} 4a^2(1 + 2\cos\theta + \cos^2\theta) \, d\theta$$
$$= 4a^2 \int_0^{\pi} \left(1 + 2\cos\theta + \frac{1 + \cos 2\theta}{2}\right) d\theta$$

(by using the identity $\cos 2\theta \equiv 2\cos^2\theta - 1$)

$$\text{area of cardioid} = 4a^2\left[\theta + 2\sin\theta + \tfrac{1}{2}\theta + \tfrac{1}{4}\sin 2\theta\right]_0^{\pi}$$
$$= 4a^2(\pi + 0 + \tfrac{\pi}{2} + 0) - 4a^2(0 + 0 + 0 + 0)$$
$$= 6\pi a^2$$

Example 6
Find the area of the finite region bounded by the half-lines $\theta = 0$ and $\theta = \frac{\pi}{3}$ and the curve with polar equation $r = a(1 + \tan\theta)$, where a is a positive constant.

As θ increases from 0 to $\frac{\pi}{3}$, r increases from a to $a(1 + \sqrt{3})$. So the curve is easily sketched and the region whose area is required is shown shaded.

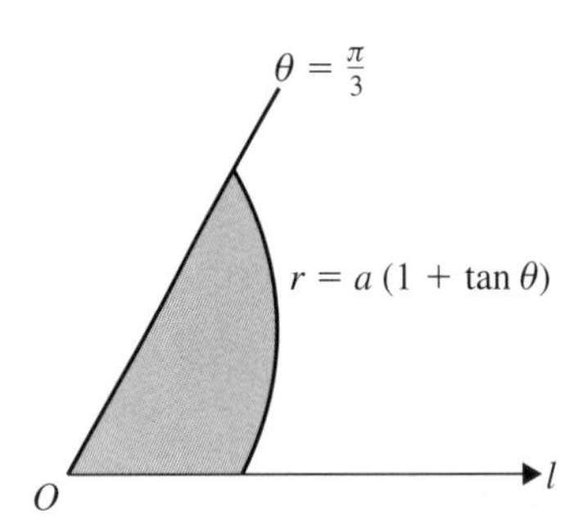

$$\text{Area of region} = \tfrac{1}{2}\int_0^{\frac{\pi}{3}} r^2 \, d\theta$$
$$= \tfrac{1}{2}\int_0^{\frac{\pi}{3}} a^2(1 + \tan\theta)^2 \, d\theta$$
$$= \frac{a^2}{2}\int_0^{\frac{\pi}{3}} (1 + 2\tan\theta + \tan^2\theta) \, d\theta$$

But $1 + \tan^2\theta \equiv \sec^2\theta$

So: $$\text{area of region} = \frac{a^2}{2}\int_0^{\frac{\pi}{3}} (2\tan\theta + \sec^2\theta)\,\mathrm{d}\theta$$

$$= \frac{a^2}{2}\Big[2\ln(\sec\theta) + \tan\theta\Big]_0^{\frac{\pi}{3}}$$

$$= \frac{a^2}{2}\left[2\ln(\sec\tfrac{\pi}{3}) + \tan\tfrac{\pi}{3} - 0 - 0\right]$$

$$= \frac{a^2}{2}(2\ln 2 + \sqrt{3})$$

Here is an example where the restriction about θ lying in an interval such as $[0, 2\pi)$ is lifted.

Example 7

Calculate the *total* area swept out by the radius vector for the spiral with polar equation $r = ae^{\theta}$, where a is a positive constant, from $\theta = 0$ to $\theta = 4\pi$.

As θ increases from 0 to 4π, r steadily increases from a to $ae^{4\pi}$ and a sketch of the spiral looks like this:

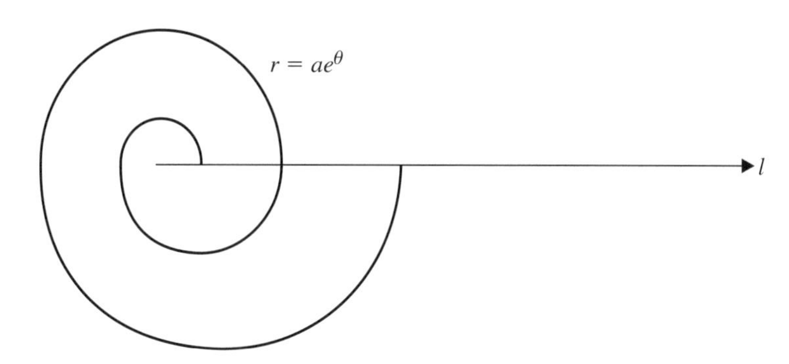

$$\text{Area swept out by radius vector} = \tfrac{1}{2}\int_0^{4\pi} r^2\,\mathrm{d}\theta$$

$$= \tfrac{1}{2}\int_0^{4\pi} a^2\mathrm{e}^{2\theta}\,\mathrm{d}\theta$$

$$= \tfrac{1}{2}a^2\int_0^{4\pi} \mathrm{e}^{2\theta}\,\mathrm{d}\theta$$

$$= \tfrac{1}{2}a^2\left[\tfrac{1}{2}\mathrm{e}^{2\theta}\right]_0^{4\pi}$$

$$= \tfrac{1}{4}a^2(\mathrm{e}^{8\pi} - 1)$$

7.5 Tangents parallel to and at right angles to the initial line l

Tangents to a polar curve that are parallel to the initial line l will occur at positions where y takes a maximum or a minimum value. **That is, then, when $r \sin \theta$ takes a maximum or a minimum value.**

Similarly, tangents to a polar curve that are at right angles to the initial line l occur at positions where x takes a maximum or a minimum value.
That is, then, when $r \cos \theta$ takes a maximum or a minimum value.

Example 8

Find, for the cardioid with equation $r = 2a(1 + \cos\theta)$, those points on the curve where
(a) tangents to the curve are parallel to l
(b) tangents to the curve are at right angles to l.

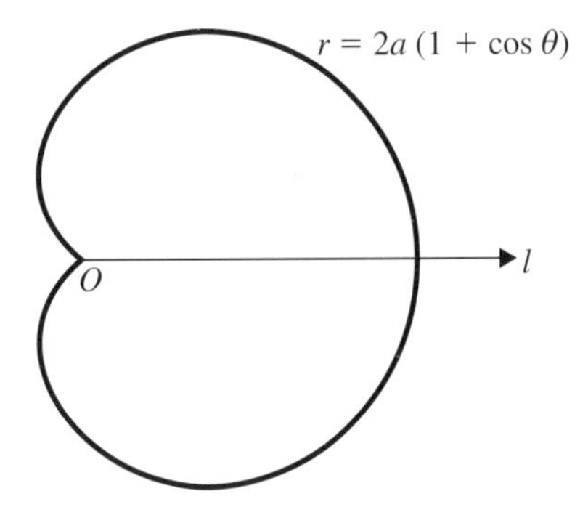

(a) Tangents are parallel to l when $\dfrac{\mathrm{d}}{\mathrm{d}\theta}(r\sin\theta) = 0$

That is $\quad \dfrac{\mathrm{d}}{\mathrm{d}\theta}(2a\sin\theta + 2a\sin\theta\cos\theta) = 0$

or: $\quad 2a(\cos\theta + \cos^2\theta - \sin^2\theta) = 0$

Since $a \neq 0$, then: $\quad \cos\theta + \cos^2\theta - \sin^2\theta = 0$

Using $\sin^2\theta \equiv 1 - \cos^2\theta$, you get

$$2\cos^2\theta + \cos\theta - 1 = 0$$

$$(2\cos\theta - 1)(\cos + 1) = 0$$

Hence $\cos\theta = \frac{1}{2}$ or $\cos\theta = -1$.

So $\theta = \pm\frac{\pi}{3}$, π and $r = 3a$, $3a$ and 0.

Tangents that are parallel to l on the curve $r = 2a(1 + \cos\theta)$ occur at the points whose polar coordinates are $(3a, \frac{\pi}{3})$, $(3a, -\frac{\pi}{3})$ and $(0, \pi)$.

(b) Tangents are at right angles to l when $\frac{\mathrm{d}}{\mathrm{d}\theta}(r\cos\theta) = 0$

That is $$\frac{\mathrm{d}}{\mathrm{d}\theta}(2a\cos\theta + 2a\cos^2\theta) = 0$$

or: $$-2a(\sin\theta + 2\cos\theta\sin\theta) = 0$$

Since $a \neq 0$, then: $$-\sin\theta - 2\cos\theta\sin\theta = 0$$

$$\Rightarrow \quad -\sin\theta(1 + 2\cos\theta) = 0$$

That is: $$\sin\theta = 0 \quad \text{or} \quad \cos\theta = -\tfrac{1}{2}$$

$$\Rightarrow \quad \theta = 0, \pi \quad \text{or} \quad \theta = \pm\tfrac{2}{3}\pi$$

and $$r = 4a, \quad \text{or} \quad r = a, a$$

Tangents are at right angles to l on the curve at the three points $(4a, 0)$, $(a, \frac{2}{3}\pi)$, $(a, -\frac{2}{3}\pi)$. Notice that at $O(0, \pi)$ both $\frac{\mathrm{d}}{\mathrm{d}\theta}(r\sin\theta)$ and $\frac{\mathrm{d}}{\mathrm{d}\theta}(r\cos\theta)$ are zero, but that, strictly speaking, the initial line is the only tangent to the curve at this point.

Exercise 7B

In questions **1–10**, the finite region R is bounded by an arc of the curve with equation $r = \mathrm{f}(\theta)$ and the half-lines with equations $\theta = \theta_1$ and $\theta = \theta_2$. Find the area of R, given that $a > 0$.

1 $\mathrm{f}(\theta) \equiv a(1 + \theta)$, $\theta_1 = -1$, $\theta_2 = 2$

2 $\mathrm{f}(\theta) \equiv a\sin\theta$, $\theta_1 = 0$, $\theta_2 = \pi$

3 $\mathrm{f}(\theta) \equiv a\tan\frac{1}{2}\theta$, $\theta_1 = 0$, $\theta_2 = \frac{\pi}{2}$

4 $\mathrm{f}(\theta) \equiv 4a\cos 2\theta$, $\theta_1 = -\frac{\pi}{4}$, $\theta_2 = \frac{\pi}{4}$

5 $\mathrm{f}(\theta) \equiv a(1 - \cos\theta)$, $\theta_1 = 0$, $\theta_2 = \pi$

6 $\mathrm{f}(\theta) \equiv a\cos 3\theta$, $\theta_1 = -\frac{\pi}{6}$, $\theta_2 = \frac{\pi}{6}$

7 $\mathrm{f}(\theta) \equiv a\sin^{\frac{1}{2}} 2\theta$, $\theta_1 = 0$, $\theta_2 = \frac{\pi}{2}$

8 $\mathrm{f}(\theta) \equiv \dfrac{a}{1 + \cos\theta}$, $\theta_1 = 0$, $\theta_2 = \frac{\pi}{2}$

9 $\mathrm{f}(\theta) \equiv a\tan\theta$, $\theta_1 = 0$, $\theta_2 = \frac{\pi}{4}$

10 $\mathrm{f}(\theta) \equiv a(3 + 2\cos\theta)$, $\theta_1 = 0$, $\theta_2 = \pi$

11 Sketch the curve given in polar coordinates by the equation

$$r = 2 + \cos\theta$$

(a) Find the area of the region enclosed by the curve.

(b) Giving answers to 2 decimal places, find the coordinates of the points on the curve where the tangent is parallel to the initial line.

12 Sketch the curve with polar equation $r = a(1 + \cos\theta)$. Show that the two tangents parallel to the initial line make with the two tangents perpendicular to the initial line a rectangle of area $\dfrac{27a^2\sqrt{3}}{8}$.

13 Sketch the spiral with equation $r = e^{\theta}$ for $0 \leqslant \theta \leqslant 2\pi$. Tangents to the curve at the points A and B are perpendicular to the initial line. Find the polar coordinates of A and B.

14 The curve C has polar equation

$$r^2 \cos 2\theta = a^2, \quad -\tfrac{\pi}{4} < \theta < \tfrac{\pi}{4}$$

Find in terms of the positive constant a

(a) the area of the region bounded by the half-lines $\theta = \pm\frac{\pi}{6}$ and the arc of the curve C in the interval $-\frac{\pi}{6} \leqslant \theta \leqslant \frac{\pi}{6}$

(b) the equation of the tangent to the curve which is perpendicular to the initial line, giving your answer in polar form.

15 The curve C has polar equation $r = a \sin\theta \sin 2\theta$, where a is a positive constant and $0 \leqslant \theta \leqslant \frac{\pi}{2}$.

(a) Find, in terms of a, the area of the region enclosed by C.

(b) Show that the tangent at the point $(\frac{3}{4}a, \frac{\pi}{3})$ is parallel to the initial line.

SUMMARY OF KEY POINTS

1 For the curve with polar equation $r = \mathrm{f}(\theta)$,

$$\text{area of shaded region} = \tfrac{1}{2}\int_{\alpha}^{\beta} r^2 \, \mathrm{d}\theta = \tfrac{1}{2}\int_{\alpha}^{\beta} [\mathrm{f}(\theta)]^2 \, \mathrm{d}\theta$$

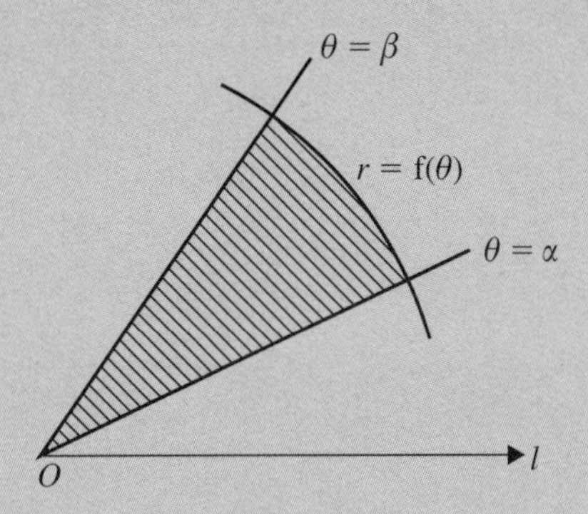

2 For tangents parallel to l, $\dfrac{\mathrm{d}}{\mathrm{d}\theta}(r\sin\theta) = 0$

For tangents perpendicular to l, $\dfrac{\mathrm{d}}{\mathrm{d}\theta}(r\cos\theta) = 0$

Review exercise 2

1 Find the solution of the differential equation

$$\frac{dy}{dx} + y\cot x = \sin 2x, \quad 0 < x < \pi$$

for which $y = 1$ at $x = \frac{\pi}{4}$. [E]

2 Find the general solution of the differential equation

$$\frac{d^2y}{dx^2} - 5\frac{dy}{dx} + 6y = 5\sin x$$ [E]

3 The curve shown has polar equation

$$r = a(1 + \tfrac{1}{2}\cos\theta),\ a > 0,\ 0 < \theta \leqslant 2\pi$$

Determine the area enclosed by the curve, giving your answer in terms of a and π.

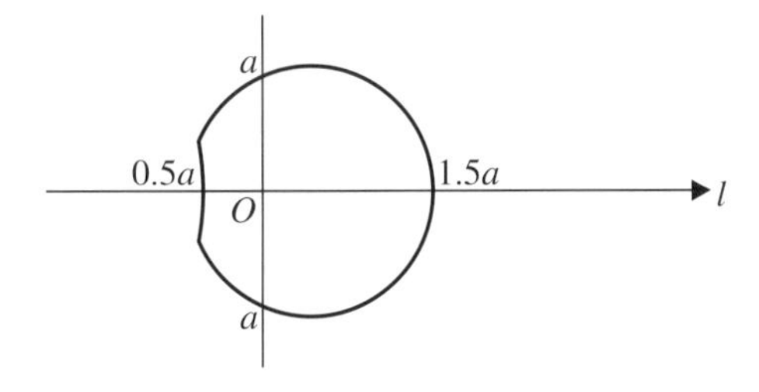

[E]

4 Find the general solution of the differential equation

$$\frac{d^2y}{dx^2} + \frac{dy}{dx} + y = 0$$ [E]

5 Given that $x > 1$ and $y > 0$, find the general solution of the differential equation

$$\frac{dy}{dx} = \frac{xy}{x-1}$$

Given further that $y = 1$ at $x = \frac{5}{3}$, find the value of y at $x = 2$, giving your answer in the form $y = ke^c$, where k and c are numbers to be found. [E]

6 Sketch the curve with equation $r = a(1 + \cos\theta)$ for $0 \leqslant \theta \leqslant \pi$ where $a > 0$. Sketch also the line with equation $r = 2a\sec\theta$ for $-\frac{\pi}{2} < \theta < \frac{\pi}{2}$ on the same diagram.
The half-line with equation $\theta = \alpha$, $0 < \alpha < \frac{\pi}{2}$, meets the curve at A *and the line with equation* $r = 2a\sec\theta$ *at* B. *If* O *is the pole, find the value of* $\cos\alpha$ *for which* $OB = 2OA$. *[E]*

7 Show that the substitution $y = z\cos x$, where z is a function of x, reduces the differential equation

$$\cos^2 x\frac{\mathrm{d}^2y}{\mathrm{d}x^2} + 2\cos x\sin x\frac{\mathrm{d}y}{\mathrm{d}x} + 2y = x\cos^3 x$$

to the differential equation

$$\frac{\mathrm{d}^2z}{\mathrm{d}x^2} + z = x$$

Hence find y given that $y = 0$ and $\frac{\mathrm{d}y}{\mathrm{d}x} = 1$ at $x = 0$. [E]

8 Find the general solution of the differential equation

$$\frac{1}{x}\frac{\mathrm{d}y}{\mathrm{d}x} + 2y = 8, \quad x > 0$$

Express your answer in the form $y = \mathrm{f}(x)$. [E]

9 Sketch, in the same diagram, the curves with equations $r = 3\cos\theta$ and $r = 1 + \cos\theta$ and find the area of the region lying inside both curves. [E]

10 Find y in terms of k and x, given that

$$\frac{\mathrm{d}^2y}{\mathrm{d}x^2} + k^2y = 0$$

where k is a constant, and $y = 1$ and $\frac{\mathrm{d}y}{\mathrm{d}x} = 1$ at $x = 0$. [E]

11 Solve the differential equation

$$\frac{\mathrm{d}y}{\mathrm{d}x} - \frac{y}{x} = x^2, \quad x > 0$$

giving your answer for y in terms of x. [E]

12 Sketch the curve C with polar equation

$$r^2 = a^2 \cos 2\theta, \quad 0 \leqslant \theta \leqslant \tfrac{\pi}{4}$$

Find the polar coordinates of the point where the tangent to C is parallel to the initial line $\theta = 0$. [E]

13 A particular solution of the differential equation

$$\frac{d^2y}{dx^2} + 4\frac{dy}{dx} + 4y = 8 \sin 2x$$

is $y = p \cos 2x + q \sin 2x$, where p and q are constants.

Find the values of p and q. Given that $y = 1$ and $\frac{dy}{dx} = 0$ at $x = 0$ find the solution of the differential equation. [E]

14 (a) Show that the substitution $v = xy$ transforms the differential equation

$$x\frac{d^2y}{dx^2} + 2(1 + 2x)\frac{dy}{dx} + 4(1 + x)y = 32e^{2x}, \quad x \neq 0$$

into the differential equation

$$\frac{d^2v}{dx^2} + 4\frac{dv}{dx} + 4v = 32e^{2x}$$

(b) Given that $v = ae^{2x}$, where a is a constant, is a particular integral of this transformed equation, find a.

(c) Find the solution of the differential equation

$$x\frac{d^2y}{dx^2} + 2(1 + 2x)\frac{dy}{dx} + 4(1 + x)y = 32e^{2x}$$

for which $y = 2e^2$ and $\frac{dy}{dx} = 0$ at $x = 1$.

(d) Determine whether or not this solution remains finite as $x \to \infty$. [E]

15 Given that the differential equation

$$\frac{d^2y}{dx^2} - 4\frac{dy}{dx} + 13y = e^{2x}$$

has a particular integral of the form ke^{2x}, determine the value of the constant k. Find the general solution of the equation. [E]

16 Shade the region C for which the polar coordinates r, θ satisfy

$$r \leqslant 4\cos 2\theta \text{ for } -\tfrac{\pi}{4} \leqslant \theta \leqslant \tfrac{\pi}{4}$$

Find the area of C. [E]

17 Use the substitution $y = vx$, where v is a function of x, to transform the differential equation

$$x^2\frac{dy}{dx} = x^2 - xy + y^2, \quad x > 0$$

into the differential equation

$$x\frac{dv}{dx} = (v-1)^2, \quad x > 0$$

Hence, given that $y = 2$ at $x = 1$, find y in terms of x. [E]

18 Sketch the curve with polar equation $r = a(1 - \cos\theta)$, where $a > 0$, stating the polar coordinates of the point on the curve at which r has its maximum value. [E]

19 Find the values of the constants p and q if $y = p\cos x + q\sin x$ satisfies the differential equation

$$\frac{d^2y}{dx^2} + 8\frac{dy}{dx} + 25y = 48\cos x - 16\sin x$$

and hence find the general solution of this differential equation. Find the solution of this differential equation for which $y = 8$ and $\frac{dy}{dx} = 3$ at $x = 0$. [E]

20 Find the general solution of the differential equation

$$\cos x\frac{dy}{dx} + 2y\sin x = \sin^2 x\cos x, \quad 0 < x < \tfrac{\pi}{2}$$ [E]

21 On the same diagram, sketch the curve C_1 with polar equation

$$r = 2\cos 2\theta, \quad -\tfrac{\pi}{4} < \theta \leqslant \tfrac{\pi}{4}$$

and the curve C_2 with polar equation $\theta = \frac{\pi}{12}$.
Find the area of the smaller region bounded by C_1 and C_2. [E]

22 Obtain the solution of

$$\frac{dy}{dx} + y\tan x = e^{2x}\cos x \quad (0 \leqslant x < \tfrac{\pi}{2})$$

for which $y = 2$ at $x = 0$, giving your answer in the form $y = f(x)$. [E]

23 Given that $x = e^t$, show that

$$\frac{d^2y}{dx^2} = e^{-2t}\left(\frac{d^2y}{dt^2} - \frac{dy}{dt}\right)$$

Hence show that the substitution $x = e^t$ transforms the differential equation

$$x^2\frac{d^2y}{dx^2} - 4x\frac{dy}{dx} + 6y = 3 \qquad (1)$$

into

$$\frac{d^2y}{dt^2} - 5\frac{dy}{dt} + 6y = 3 \qquad (2)$$

Hence find the general solution of the differential equation (1). [E]

24 Sketch on the same diagram the circle with polar equation $r = 4\cos\theta$ and the line with polar equation $r = 2\sec\theta$. State polar coordinates for their points of intersection. [E]

25 Find the general solution of the differential equation

$$\frac{d^2y}{dx^2} + 2\frac{dy}{dx} + 5y = 0$$ [E]

26 Find the general solution of the differential equation

$$\frac{d^2y}{dx^2} + 2\frac{dy}{dx} + 10y = 0$$

(a) Find the constants k and p such that ke^{px} is a particular integral of the differential equation

$$\frac{d^2y}{dx^2} + 2\frac{dy}{dx} + 10y = 54e^{2x}$$

and hence find the solution of this differential equation for which $y = 0$ and $\frac{dy}{dx} = 3$ at $x = 0$.

(b) (i) Find the solution of

$$\frac{d^2y}{dx^2} + 2\frac{dy}{dx} + 10y = 0$$

for which $y = 0$ and $\frac{dy}{dx} = 3$ at $x = 0$.

(ii) Find the stationary values of y.

(iii) Hence show that the moduli of consecutive stationary values of y form a geometric progression. [E]

27 Sketch, on the same diagram, the curves C_1, C_2 whose polar equations are

$$C_1 : r = a(1 + \cos\theta), -\pi < \theta \leqslant \pi$$
$$C_2 : r = b(1 - \cos\theta), -\pi < \theta \leqslant \pi$$

where a *and* b *are positive constants with* $b > a$.
Find the points common to C_1 and C_2.
Calculate the area of the finite region bounded by the arcs with polar equations

$$r = a(1 + \cos\theta), \tfrac{\pi}{2} \leqslant \theta \leqslant \pi$$
$$r = a(1 - \cos\theta), 0 \leqslant \theta \leqslant \tfrac{\pi}{2}$$

[E]

28 Find the general solution of the differential equation

$$\frac{dy}{dx} - y\tan x = \sec^2 x, 0 < x < \tfrac{\pi}{2}$$

giving your answer for y in terms of x. [E]

29 Solve the differential equation

$$\frac{d^2y}{dx^2} - 4\frac{dy}{dx} + 4y = 4x$$

given that $y = 3$ and $\frac{dy}{dx} = 8$ at $x = 0$.

Show that $y = 2 + 5e^2$ at $x = 1$. [E]

30 Obtain the general solution of the differential equation

$$\frac{d^2y}{dx^2} + 4\frac{dy}{dx} + 13y = e^{-3x}$$

[E]

31 Show that the substitution $z = \dfrac{1}{y^2}$ transforms the differential equation

$$\frac{\mathrm{d}y}{\mathrm{d}x} - y = 2xy^3,\ y > 0 \qquad (1)$$

into the differential equation

$$\frac{\mathrm{d}z}{\mathrm{d}x} + 2z = -4x$$

Hence find the general solution of the differential equation (1).

[E]

32 Find the solution of

$$\frac{\mathrm{d}y}{\mathrm{d}x} + \frac{y}{x} = \tfrac{1}{2}\sin\left(\tfrac{1}{2}x\right),\ x > 0$$

such that $y = 1$ at $x = 2\pi$.
Given that as $x \to 0$, $y \to k$, determine the value of k. [E]

33 Find the values of the constants a and b in order that $y = a\cos 3x + b\sin 3x$ shall satisfy the differential equation

$$\frac{\mathrm{d}^2y}{\mathrm{d}x^2} + 2\frac{\mathrm{d}y}{\mathrm{d}x} = 5\sin 3x$$

Hence find the general solution of the differential equation.

[E]

34 A curve which passes through the point $(1, 1)$ has equation $y = \mathrm{f}(x)$, where

$$\frac{\mathrm{d}y}{\mathrm{d}x} + \frac{2(x-1)}{x^2 - 2x + 2}y = \frac{2(x+1)}{x^2 - 2x + 2}$$

(a) Solve this differential equation, to show that

$$\mathrm{f}(x) \equiv \frac{x^2 + 2x - 2}{x^2 - 2x + 2}$$

(b) Show that $-1 \leqslant \mathrm{f}(x) \leqslant 3$. [E]

35 Show that, if $y = xz$, where z is a function of x, then

$$\frac{\mathrm{d}y}{\mathrm{d}x} = z + x\frac{\mathrm{d}z}{\mathrm{d}x}$$

Using the substitution $y = xz$, or otherwise, find the general solution of the differential equation

$$x\frac{\mathrm{d}y}{\mathrm{d}x} = y + 2x(x + y) \qquad \text{[E]}$$

36 Obtain the values of the constants a and b, given that $y = a\cos 2x + b\sin 2x$ is a particular solution of

$$3\frac{\mathrm{d}^2y}{\mathrm{d}x^2} - 4\frac{\mathrm{d}y}{\mathrm{d}x} - 4y = 10\cos 2x$$

Hence obtain the general solution of the differential equation.

[E]

37

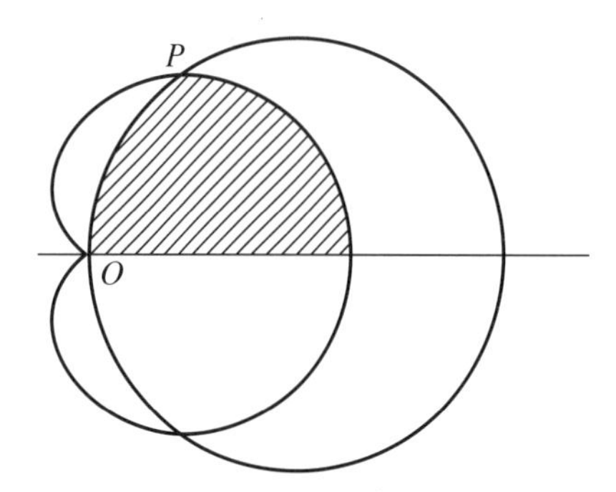

The diagram shows a sketch of the curves with polar equations

$$r = a(1 + \cos\theta) \quad \text{and} \quad r = 3a\cos\theta, \quad a > 0$$

(a) Find the polar coordinates of the point of intersection P of the two curves.

(b) Find the area, shaded in the figure, bounded by the two curves and by the initial line $\theta = 0$, giving your answer in terms of π.

[E]

38 (a) Given that $ye^{2x} = A\cos 3x + B\sin 3x$, where A and B are constants, show that

$$\frac{\mathrm{d}^2y}{\mathrm{d}x^2} + 4\frac{\mathrm{d}y}{\mathrm{d}x} + 13y = 0.$$

For the differential equation

$$\frac{\mathrm{d}^2y}{\mathrm{d}x^2} + 4\frac{\mathrm{d}y}{\mathrm{d}x} + 13y = 63e^{4x},$$

find

(b) the general solution,

(c) the specific solution for which $\frac{\mathrm{d}y}{\mathrm{d}x} = y = 0$ at $x = 0$.

39 Use the substitution $z = y^{-2}$ to transform the differential equation

$$xy - \frac{dy}{dx} = y^3 e^{-x^2}$$

into the differential equation

$$\frac{dz}{dx} + 2xz = 2e^{-x^2}$$

Hence obtain the general solution of the first differential equation. [E]

40

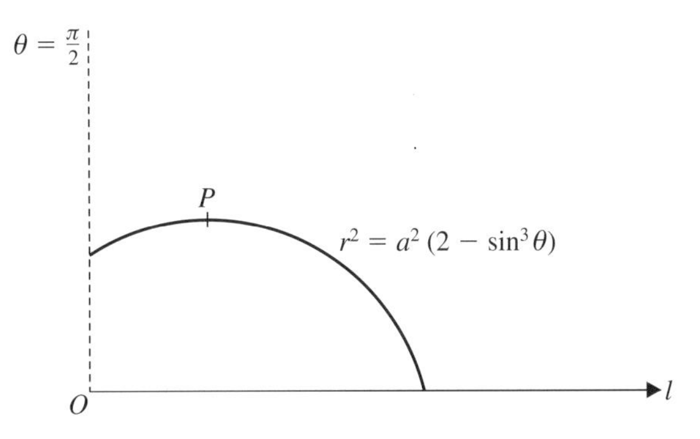

The diagram shows a sketch of the curve with equation

$$r^2 = a^2(2 - \sin^3\theta), \quad 0 \leqslant \theta \leqslant \tfrac{\pi}{2}$$

where a is a positive constant.
By considering the stationary values of $r^2 \sin^2\theta$, or otherwise, find the polar coordinates of P, where P is a point on the curve at which the tangent is parallel to the initial line l.

41 At any point $P(x, y)$ on a curve C,

$$\frac{dy}{dx} + y \cot x = \cos 3x$$

Given that C passes through the point $A\left(\frac{\pi}{2}, 0\right)$, find an equation for C in the form $y = f(x)$.

42 Obtain a cartesian equation for the curve with polar equation
(a) $r^2 = \sec 2\theta$ (b) $r^2 = \operatorname{cosec} 2\theta$

43 By differentiating twice and eliminating the arbitrary constants A and B, obtain the second order differential equation of which the general solution is

$$y = A\cos 3x + B\sin 3x + x^2 + 2$$ [E]

44 Obtain, in the form $r = \mathrm{f}(\theta)$, a polar equation of the curve whose cartesian equation is

(a) $x + 4y = 3$ (b) $x^2 + xy = 2y$

45 Find the solution of the differential equation

$$\frac{\mathrm{d}^2y}{\mathrm{d}x^2} + 3\frac{\mathrm{d}y}{\mathrm{d}x} + 2y = 2\mathrm{e}^{-x}$$

for which $\frac{\mathrm{d}y}{\mathrm{d}x} = 3$ at $x = 0$ and $y = 1$ at $x = 0$. [E]

Examination style paper

FP1

Answer all questions **Time 90 minutes**

1. $$f(x) \equiv x - \tan\left(\tfrac{11}{12}x\right)$$

(*a*) Prove that the equation $f(x) = 0$ has a root α in the interval $[0.5, 0.6]$. **(3 marks)**

(*b*) Working to four significant figures, use linear interpolation on the interval $[0.5, 0.6]$ to find an estimate for α, giving your answer to three decimal places. **(4 marks)**

2. Using the substitution $y = tx$, or otherwise, find the general solution of the differential equation

$$x\frac{dy}{dx} = x + y, \quad x > 0.$$ **(7 marks)**

3. (*a*) Sketch the curve with equation $y = |x^2 - 8x + 12|$, stating, on your sketch, the coordinates of points at which the curve meets the coordinate axes. **(3 marks)**

(*b*) Hence, or otherwise, find the set of values of x for which $|x^2 - 8x + 12| < 3$. **(7 marks)**

4. $$u_r = \frac{1}{r+2} - \frac{1}{r+4}$$

(*a*) Prove that $\displaystyle\sum_{r=1}^{n} u_r = \frac{7}{12} - \frac{2n+7}{(n+3)(n+4)}$. **(5 marks)**

(*b*) Find the exact value of $\displaystyle\sum_{r=6}^{17} u_r$, giving your answer as a single rational number in the form $\dfrac{p}{q}$, where p and q are integers. **(5 marks)**

5. The polar curve shown has equation $r = 1 + 2\cos\theta$, $-\frac{2}{3}\pi \leqslant \theta \leqslant \frac{2}{3}\pi$. The pole is at O and the initial line is l. At the points A and B on the curve, the tangent to the curve is at right-angles to l. Find

(*a*) the area of the region enclosed by the curve. **(6 marks)**

(*b*) the polar coordinates of A and B. **(6 marks)**

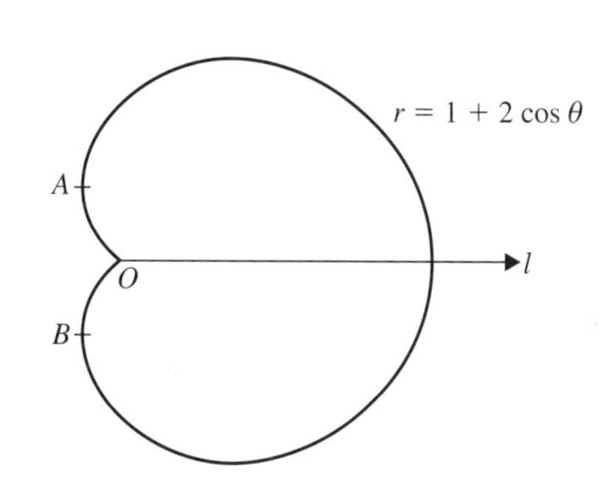

6. (*a*) One root of the equation $z^3 - 3z - 52 = 0$ is $-2 + 3\mathrm{i}$.

(i) Find the other two roots.

(ii) Verify that the sum of all three roots is zero.

(iii) Display the roots on an Argand diagram. **(9 marks)**

(*b*) Find the real numbers p and q for which

$$(5 - 4\mathrm{i})(3 + p\mathrm{i}) = q - 2\mathrm{i}.$$ **(5 marks)**

7. Prove that the substitution $x = \mathrm{e}^t$ transforms the differential equation

$$ax^2\frac{\mathrm{d}^2y}{\mathrm{d}x^2} + bx\frac{\mathrm{d}y}{\mathrm{d}x} + cy = 0,$$

where a, b and c are constants, into the differential equation

$$a\frac{\mathrm{d}^2y}{\mathrm{d}t^2} + (b - a)\frac{\mathrm{d}y}{\mathrm{d}t} + cy = 0.$$ **(8 marks)**

Hence find the general solution of the differential equation

$$x^2\frac{\mathrm{d}^2y}{\mathrm{d}x^2} + x\frac{\mathrm{d}y}{\mathrm{d}x} + 4y = \ln x, \quad x > 0$$ **(7 marks)**

Answers

Edexcel accepts no responsibility whatsoever for the accuracy or method of working in the answers given for examination questions.

Exercise 1A

1 $x > 5\frac{1}{2}$

2 $x \geqslant 24$

3 $x < 1$ or $x > 5$

4 $-3\frac{1}{2} \leqslant x \leqslant -3$

5 $-1 < x < 4$

6 $1 < x < 4$

7 $0 < x < 1$

8 $x < 1$ or $x > 2$

9 $-3 < x < -1$

10 $-\frac{2}{3} < x < 1$

11 $x \geqslant 3$ or $-2 \leqslant x \leqslant 0$

12 $x > 5$ or $0 < x < 2$

13 $x > -1$

14 $0 < x < 1, x < -1$

15 $x > 1$ or $x < -11$

16 $x \geqslant 3$ or $-2 \leqslant x \leqslant -1$

17 $2 < x < 8$ or $x < -1$

19 $x < -3$ or $-1 < x < 1$

20 $x \geqslant 0$

Exercise 1B

1 $-3 < x < 3$

2 $x > 2$ or $x < -2$

3 $x < -1$ or $0 < x < 5$

4 $x < -2$ or $x > \frac{2}{3}$

5 $1 < x < 3 - \sqrt{2}$ or $3 + \sqrt{2} < x < 5$

6 $x < -3$ or $-\frac{1}{2} < x < \frac{1}{2}$

7 $x < -1$ or $0 < x < 1$

8 $0 < x < \frac{1}{2}$ or $x > \frac{5}{8}$

9 $-2 < x < 0$ or $x > 1$

10 $x < -3$ or $-1 < x < 1$

Exercise 1C

1 $x < -2$ or $x > 6$

2 $-5 < x < 2$

3 $x < \frac{2}{5}$

4 $x \geqslant -\frac{1}{2}$ or $x \leqslant -2$

5 $1 \leqslant x \leqslant \sqrt{7}$ or $-\sqrt{7} \leqslant x < -1$

6 $-2 \leqslant x \leqslant 3$

7 $-2 < x < \frac{2}{3}$

8 $1 < x < 3$

9 $-1 - \sqrt{17} < x < -2$
or $0 < x < -1 + \sqrt{17}$

10 $x < 2$

12 $-4 \leqslant x \leqslant 0$ or $x \geqslant \frac{1}{4}$

Exercise 2A

1 $1 - \dfrac{1}{n+1}$

2 $n^2 + 2n$

3 $\dfrac{n}{2n+1}$

4 $n^2(n+1)$

5 $\dfrac{n}{n+1}$

6 $\frac{1}{4}n(n+1)(n+2)(n+3)$

7 $\dfrac{n(n+3)}{4(n+1)(n+2)}$

8 $\dfrac{n(n+2)}{(n+1)^2}$

9 $\frac{1}{3}n(n+1)(n+2)$

10 $1 - \dfrac{1}{(n+1)!}$

Exercise 2B

1 819

2 4320

3 4760

4 3230

5 $\frac{20}{21}$

6 29 141

7 2.95 (3 s.f.)

8 0.1655 (4 d.p.)

9 130 663

10 89 670

13 $\frac{20}{41}$

14 $\dfrac{n}{6}(n+1)(14n+1)$

15 $\frac{28}{1755}$

17 6734 **18** −2 010 000

19 $-n(2n+1)$

20 $\frac{n}{6}(n+1)(4n+5)+2^{n+3}-8$

Exercise 3A

1 (a) 8i (b) $i\sqrt{7}$ (c) 4 − 9i
(d) 3 − 5i (e) 3i

2 (a) −i (b) −i (c) −i
(d) −5 (e) −5 + 5i

3 (a) 4 + 10i (b) 3 + 4i
(c) 404 − 1121i (d) 15 − 22i
(e) −6 + 35i

4 (a) 2 − 4i (b) 3 + 6i (c) −5 − 2i
(d) −7 + 3i (e) −4 − 2i (f) 6
(g) −3i (h) 7 + 3i

5 (a) 6 − 4i (b) −9 − 2i
(c) −5 − 13i (d) −1 + 10i
(e) 4 + i (f) −1 + 3i
(g) −2 + 2i

6 (a) 7 + i (b) −34 + 13i
(c) −23 + 14i (d) −24 − 10i
(e) 2 − 11i (f) 8 + 6i
(g) 17 − i

7 (a) $-\frac{1}{5}(12+11i)$ (b) $\frac{1}{10}(1+7i)$
(c) $\frac{1}{25}(11+2i)$ (d) $\frac{1}{5}(1-2i)$
(e) $\frac{1}{13}(-5+12i)$ (f) $-\frac{1}{5}(7+4i)$
(g) $-\frac{6}{25}(3+4i)$ (h) $\frac{1}{25}(4+3i)$

8 (a) ±5i (b) 0, ± 8i (c) 2 ± i
(d) −3 ± i (e) 2 ± 5i
(f) $\frac{1}{4}(-3 \pm i\sqrt{47})$
(g) $\frac{1}{3}(-1 \pm i\sqrt{2})$ (h) $\frac{1}{3}(1 \pm i\sqrt{5})$

9 (a) $\frac{2}{5}$ (b) $\frac{1}{130}(47-i)$ (c) $\frac{4}{5}(7-4i)$

10 $-\frac{6}{5}(1-2i)$

12 $\lambda = \frac{1}{2}$; $-\frac{1}{4}$

Exercise 3B

2 (a) $|z| = \sqrt{13}$, $\arg z = -0.588^c$
(b) $|z| = \sqrt{10}$, $\arg z = 0.322^c$
(c) $|z| = 6$, $\arg z = 1.57^c$
(d) $|z| = 5$, $\arg z = 3.14^c$
(e) $|z| = \sqrt{5}$, $\arg z = 2.68^c$
(f) $|z| = \sqrt{10}$, $\arg z = -1.25^c$
(g) $|z| = 2$, $\arg z = 1.05^c$
(h) $|z| = 13$, $\arg z = 1.97^c$
(i) $|z| = \frac{5}{2}$, $\arg z = 1.05^c$
(j) $|z| = \frac{\sqrt{6}}{3}$, $\arg z = -0.421^c$
(k) $|z| = \sqrt{65}$, $\arg z = -0.124^c$
(l) $|z| = \frac{1}{5}\sqrt{10}$, $\arg z = 1.25^c$

3 (a) 5, 13, $\sqrt{2}$, 25
(b) 53.1°, −67.4°, −135.0°, 106.3°

4 (a) $z = 1 + i$, $|z| = \sqrt{2}$, $\arg z = 0.79^c$
(b) $z = 3(2-i)$, $|z| = 3\sqrt{5}$, $\arg z = -0.46^c$
(c) $z = -4$, $|z| = 4$, $\arg z = 3.14^c$

5 $z^2 = 3 - 4i$, $z^3 = 2 - 11i$
$|z| = \sqrt{5}$, $|z^2| = 5$, $|z^3| = 5\sqrt{5}$
$\arg z = -0.46^c$, $\arg z^2 = -0.93^c$
$\arg z^3 = -1.39^c$

6 (a) $|z_1| = 1$, $\arg z_1 = \frac{\pi}{2}$
(b) $|z_2| = 1$, $\arg z_2 = \frac{\pi}{4}$
(c) $|z_3| = 1$, $\arg z_3 = \pi$

7 $5 + 12i$; $\frac{1}{169}(5-12i)$; 13; $\frac{1}{13}$; 1.18^c; -1.18^c

9 $-\frac{6}{5}(1-2i)$; $\frac{6\sqrt{5}}{5}$

10 (a) (i) $\sqrt{52}$ (ii) -3.08^c
(b) (i) $\frac{1}{2}\sqrt{13}$ (ii) 1.11^c
(c) (i) $(277+156\sqrt{3})^{\frac{1}{2}}$ (ii) -2.73^c

11 $-(1+i)$

13 (a) $8(-1+i\sqrt{3})$
(b) $\frac{1}{8}(1-i\sqrt{3})$
(c) $2(\sqrt{3}-i)$

14 (b) 127°

15 (a) $\frac{1}{6}$
(b) 2.03^c

16 (a) 138.0°
(b) 2 − i
(c) 2.2

17 $z^2 = -\frac{1}{2}(1+i\sqrt{3})$

18 $k = -3$

19 $|z| = 1$, $\arg z = -0.52^c$, $z^2 = \frac{1}{2}(1-i\sqrt{3})$

20 (a) 56 − 33i; $\frac{1}{25}(16+63i)$

21 $|z_1| = 2$; $\arg z_1 = -\frac{\pi}{6}$
$|z_2| = 2$; $\arg z_2 = \frac{3\pi}{4}$
$\frac{z_2}{z_1} = \frac{1}{4}\left[-\sqrt{6} - \sqrt{2} + i(\sqrt{6} - \sqrt{2})\right]$

22 $\arg z_1 = \frac{2\pi}{3}$, $\arg z_2 = \frac{\pi}{6}$; $\frac{z_1}{z_2} = i$; $\arg\frac{z_1}{z_2} = \frac{\pi}{2}$

23 $z_1 = 2(1 - i)$; $z_2 = -2$
(b) $-\frac{\pi}{2}$

24 $|z_1| = \sqrt{2}$, $\arg z_1 = \frac{\pi}{4}$
$|z_2| = 2$, $\arg z_2 = -\frac{\pi}{6}$

25 (b) $|z_1| = 2\sqrt{2}$, $|z_2| = \sqrt{10}$, $PQ = \sqrt{2}$
(d) $-1 + i$

26 (a) $z = 4 - 3i$; $w = 3 + 4i$
(c) 90°

Exercise 3C

1 (a) $5\left(\cos\frac{\pi}{2} + i\sin\frac{\pi}{2}\right)$
(b) $7(\cos 0 + i\sin 0)$
(c) $3\left[\cos\left(-\frac{\pi}{2}\right) + i\sin\left(-\frac{\pi}{2}\right)\right]$
(d) $6(\cos\pi + i\sin\pi)$
(e) $2\left(\cos\frac{\pi}{3} + i\sin\frac{\pi}{3}\right)$
(f) $6\left[\cos\left(-\frac{\pi}{6}\right) + i\sin\left(-\frac{\pi}{6}\right)\right]$
(g) $5(\cos 2.21 + i\sin 2.21)$
(h) $\sqrt{2}\left[\cos\left(-\frac{\pi}{4}\right) + i\sin\left(-\frac{\pi}{4}\right)\right]$
(i) $10[\cos(-0.927) + i\sin(-0.927)]$
(j) $\cos\frac{\pi}{3} + i\sin\frac{\pi}{3}$
(k) $4\left(\cos\frac{\pi}{6} + i\sin\frac{\pi}{6}\right)$
(l) $\frac{\sqrt{221}}{17}[\cos(-1.91) + i\sin(-1.91)]$

2 (a) $\frac{3}{2}(1 + i\sqrt{3})$ (b) $-\frac{5}{\sqrt{2}} + \frac{5i}{\sqrt{2}}$
(c) $3\sqrt{3} - 3i$ (d) $4i$
(e) -10 (f) $-\frac{9}{2}(\sqrt{3} + i)$
(g) $\frac{7}{3\sqrt{2}}(1 + i)$ (h) $\frac{3}{\sqrt{2}}(-1 + i)$
(i) $4(-\sqrt{3} + i)$ (j) $\frac{2}{3}(\sqrt{3} + i)$

3 $\cos(-\pi) + i\sin(-\pi) = -1$

Exercise 3D

1 (a) $\pm(3 + 2i)$ (b) $\pm(4 - 3i)$
(c) $\pm(2 - i)$ (d) $\pm\sqrt{10}(1 - i)$
(e) $\pm(2 - i\sqrt{3})$

2 (a) $(-4, 4)$, $(16, -1)$
(b) $(7, 4)$
(c) $(1, -2)$
(d) $\left(-\frac{650}{49}, -\frac{50}{49}\right)$ (e) $(2, 4)$

3 (a) $A = \frac{7}{5}$, $B = -\frac{4}{5}$
(b) $A = 2 - i$, $B = 2 + i$

4 $x = 1$, $y = -1$

5 (a) $p = 3$, $q = 1$
(b) $p = -\frac{2}{5}$, $q = \frac{6}{5}$

6 $x = -\frac{7}{25}$, $y = -\frac{24}{25}$

7 $x = \frac{2}{13}$, $y = \frac{3}{13}$

8 $x = 1$, $y = -\frac{1}{2}$

9 $a = 0$, $b = \pm 2$

10 $\frac{1}{z} = \frac{1}{169}(5 + 12i)$
$z^{\frac{1}{2}} = \pm(3 - 2i)$

11 $x = 2$, $y = 0$

12 (a) $\frac{1}{50}(147 - 71i)$
(b) $m = -2$, $n = 3$

13 $x = 3$, $y = -1$
(a) $\sqrt{10}$
(b) -18.4° (-0.322^c)

14 (a) (iii) -0.49^c
(b) $w = \pm(1 + 2i)$

15 $A = 27$, $B = 8$

16 $\pm i$, ± 1

17 $2 - 3i$, $\frac{1}{2}$

18 $1 + 2i$, $\frac{1}{2}$

19 $-2i$, -4, 1

20 $\pm i$, $\pm 2i$

21 $\frac{1}{2}(1 \pm i\sqrt{7})$, 2

22 $-ai$, b

Exercise 4A

1 2.49 **2** 0.80 **3** 0.69
4 1.24 **5** 0.81
6 $x_0 = 0.933\,96\ldots$; 0.934 **7** 0.52
8 0.669 **9** 0.199 936 **10** 3.20
11 1.38 **12** 2.24 **13** −1.83
14 0.587 **15** 1.54

16 (a) 0.17 (b) 2.04 (c) 3.11 (d) 1.31
(e) 4.33 (f) 0.51 (g) 3.59

17 3.73 **18** 0.5

19 (a) 2.207 99 ... (b) 2.207 93 ...

20 0.905 08 ...

21 (a) 0.817 886 ... (b) 0.809 86 ...

22 0.357

23 (b) −0.75 (c) −0.82

24 (b) 1.15

Review exercise 1

1 (a) $z_1 z_2 = 7 + 4\text{i}$, $\frac{z_1}{z_2} = \frac{1}{5} + \frac{8}{5}\text{i}$

2 1.86

3 $x < 4$, $x > 4\frac{2}{5}$

4 $\frac{1}{12}n(n+1)(n+2)(3n+17)$

5 (a) 42 925 (b) 112 761 (c) 85 320

7 (b) 1.47

8 $-2 < t < 1$

9 modulus = 1, argument = $-\frac{\pi}{3}$

10 (a) 3 (b) 3.43

11 $-\frac{5\pi}{6} < x < -\frac{\pi}{6}$, $\frac{\pi}{6} < x < \frac{5\pi}{6}$, $\frac{7\pi}{6} < x < \frac{11\pi}{6}$

12 $z = 3 + \text{i}$

13 26 800

14 $-2 < x < 3$ or $x > 8$

15 (b) 1.4 (c) 1.55

16 (a) 2.45 (b) $p = 3$, $q = -1$ (c) $3 + \text{i}$

17 (b) $|z| = \frac{1}{\sqrt{2}}$, $\arg z = -\frac{3}{4}\pi$

18 1.444

19 (b) $x < \frac{8}{9}$

21 (b) 1.53 is correct to 2 d.p.

22 $-6 < x < 0$, $x > \frac{1}{2}$

23 $\frac{1}{2x-1} - \frac{1}{2x+1}$

24 (a) $\sqrt{2}$ (b) 5
(c) $-2 + 3\text{i}$ (d) 123.7°

25 (b) 2.79
(c) correct to 2 d.p.
(d) 0.146, 3 positive roots

26 5, −3, $1 - 2\text{i}$

27 $-2 < x < -1$, $0 < x < 2$

28 $\frac{n(n+2)}{(n+1)^2}$

29 $x_2 = \frac{3}{2}$, $x_3 = \frac{17}{12}$; error 2×10^{-6}

30 $(z-3)(z+1)(z^2 - 6z + 34)$; 3, −1, $3 + 5\text{i}$

31 $x < 0$ or $1 < x < 4$

32 (b) $\frac{1}{x+1} - \frac{1}{x+2}$

33 (b) 4, −2
(c) $-\frac{\pi}{4}$

34 (c) 0.8604

35 (a) 2, $2 - \text{i}$
(b) −10

36 $x < -2$, $0 < x < 1$

37 $\frac{n}{6}(n+1)(2n+7)\ln 2$

38 $-\frac{5\sqrt{3}}{4} + \frac{3}{4}\text{i}$

39 $x < -3$ or $-1 < x < 1$

40 0.433

41 (a) 2 (b) 150° (c) −60°

42 2, $2 - 3\text{i}$

43 2.08

45 $-3 < x < 1$

46 (a) $\frac{3 - 4\text{i}}{25}$ (b) $\frac{-7 - 24\text{i}}{625}$, −106.3°

47 3.12

48 (a) (ii) $\frac{3}{4}\pi$ (b) $\pm(5 - 3\text{i})$

49 1.78

50 (b) $x < a$ or $x > a + \frac{1}{2}$

51 $\frac{2}{r-1} - \frac{1}{r} - \frac{1}{r+1}$

52 1.896

53 (a) $\frac{7}{3}$ (b) $p = -3$, $q = -4$

54 (a) (i) $3 - 7\text{i}$ (ii) −16.6°
(b) (i) $|2w\text{i}| = 2|w|$
(ii) $\arg(2w\text{i}) - \arg w = 90°$

55 (b) 1.15, 1.11

56 $z_1 z_2 = -1 + 2\text{i}$, $\frac{z_1}{z_2} = \frac{11}{5} + \frac{2}{5}\text{i}$, $z_3 = \frac{6}{5} + \frac{12}{5}\text{i}$

57 3.65

58 2.605

59 (b) $5\sqrt{2}$
(c) (i) $-1 - \text{i}$ (ii) $-\frac{3}{4}\pi$

60 0.92

61 (b) 1
(d) 0.567

62 Roots are $1 \pm 2\text{i}$, $\frac{1}{2}$ and $p = 12$

63 (a) $11 + 7\text{i}$ (b) $3 + \text{i}$

64 z_1 mod. 1, arg $\frac{\pi}{2}$
z_2 mod. 1, arg $\frac{\pi}{4}$

65 0.80, 1.38

66 (b) $1 + 3i$ and 3; $\lambda = 30$

Exercise 5A

1 $2y = \sin 2x + C$

2 $3\cos\frac{1}{3}y + x + C = 0$

3 $\sec y = C\sin x$

4 $e^{-2y} + 2\tan x = C$

5 $\ln\left|\frac{y-1}{y+1}\right| + 4e^{-\frac{1}{2}x} = C$

6 $x - y^{-1} - x\ln x = C$

7 $2\ln|y| - e^{x^2} = C$

8 $e^x + e^{-y} = C$

9 $Cy^2 = \frac{x-1}{x+1}$

10 $\frac{1}{x} + \cot y = C$

11 $\frac{1}{y} + 4x + 6 = 0$

12 $\ln|y| = e^x - 1$

13 $y = \tan x - x + \frac{\pi}{4} - 1$

14 $2e^{3x} = 2e + 3e^{-1} - 3e^{-2y}$

15 $y = 4x,\ x > 0$

16 $2y^{\frac{1}{2}} = 5 - e^{-x}$

17 $1 - e^{-y} = \ln\left|\tan\frac{x}{2}\right|$

18 $\ln|\sin y| = 3\ln|\sin x| + x - \ln 2 - \frac{\pi}{2}$

19 $y = -\frac{40}{3}\ln\left|\frac{5 - 3\sin x}{8}\right|$

20 $\frac{y+1}{y} = \frac{3}{4}(1 + \cos^2 x)$

21 $y^2 = 2x^2 + 1$

22 $\frac{x^2}{2} = \ln\left|\frac{2y}{y+1}\right|$

23 $\tan x = 2\sin y$

24 $\frac{1}{y+2} + \frac{1}{2}e^{x^2} = 1$

Exercise 5B

1 $y = 4x + C$

2 $y^2 = 8x + C$

3 $y = Cx$

4 $y = \frac{1}{2}e^{2x} + C$

5 $y = \sin x + C$

6 $y^2 = 2x - x^2 + C$

7 $2y^2 = x^2 + C$

8 $xy = C,\ x > 0,\ y > 0$

Exercise 5C

1 $3xy = x^3 + C$

2 $4xy^2 = x^4 + C$

3 $y\sin x = \ln|\sec x| + C$

4 $ye^{2x} = \sin x - x\cos x + C$

5 $2\frac{x}{y} = \sin 2x + C$

6 $xy = x\sin x + \cos x + C$

7 $x^2y = x^4 + x^3 + C$

8 $\frac{y}{x} = \frac{1}{2}(\ln|x|)^2 + C$

9 $yx^{\frac{1}{2}} = C - \frac{1}{2}x^2$

10 $y\sin x = \frac{3}{2}\cos^2 x - \cos^4 x + C$

11 $y\sec^2 x = \sec x + C$

12 $\frac{y}{1+x} = x - \ln|1+x| + C$

13 $y = x^3\ln|x| - x^3 + 3x^2$

14 $y = \frac{1}{5}(2\sin x - \cos x + e^{-2x})$

15 $y\cot^2\frac{x}{2} = 2\ln\left|\sin\frac{x}{2}\right| + C$

16 $ye^{2x} = \frac{x^4}{4} + \ln|x| + C$;
$y = \frac{1}{4}(x^4 - 1)e^{-2x} + e^{-2x}\ln x$

17 $y = [e^x(x^2 - 2x + 2) + 1 - e]x^{-3}$

18 $y = \frac{5x}{2} - \frac{1}{2x}$

Exercise 6A

1 $y = Ae^x + Be^{2x}$

2 $y = Ae^{-x} + Be^{-3x}$

3 $y = Ae^x + Be^{4x}$

4 $y = Ae^{2x} + Be^{-9x}$

5 $y = Ae^{4x} + Be^{-2x}$

6 $y = Ae^{-2x} + Be^{-3x}$

7 $y = Ae^{2x} + Be^{-\frac{2}{3}x}$

8 $y = e^x(Ae^{x\sqrt{3}} + Be^{-x\sqrt{3}})$

9 $y = Ae^{\frac{2}{3}x} + Be^{-\frac{3}{2}x}$

10 $y = Ae^{3x} + Be^{-\frac{7}{3}x}$

Exercise 6B

1 $y = (A + Bx)e^{x}$
2 $y = (A + Bx)e^{-2x}$
3 $y = (A + Bx)e^{3x}$
4 $y = (A + Bx)e^{-4x}$
5 $y = (A + Bx)e^{-\frac{1}{2}x}$
6 $y = (A + Bx)e^{\frac{1}{3}x}$
7 $y = (A + Bx)e^{\frac{3x}{2}}$
8 $y = (A + Bx)e^{-\frac{5}{3}x}$
9 $y = (A + Bx)e^{x\sqrt{2}}$
10 $y = (A + Bx)e^{-x\sqrt{\frac{5}{2}}}$

Exercise 6C

1 $y = A\cos x + B\sin x$
2 $y = A\cos 5x + B\sin 5x$
3 $y = A\cos\dfrac{3x}{2} + B\sin\dfrac{3x}{2}$
4 $y = A\cos\dfrac{7x}{4} + B\sin\dfrac{7x}{4}$
5 $y = e^{x}(A\cos 2x + B\sin 2x)$
6 $y = e^{-2x}(A\cos x + B\sin x)$
7 $y = e^{3x}(A\cos x + B\sin x)$
8 $y = e^{-4x}(A\cos 3x + B\sin 3x)$
9 $y = e^{\frac{1}{2}x}(A\cos x + B\sin x)$
10 $y = e^{\frac{2x}{5}}\left(A\cos\dfrac{3x}{5} + B\sin\dfrac{3x}{5}\right)$

Exercise 6D

1 $y = Ae^{x} + Be^{3x} + 4$
2 $y = Ae^{-x} + Be^{-2x} + 2x - 3$
3 $y = (A + Bx)e^{x} + e^{2x}$
4 $y = (A + Bx)e^{-2x} + \frac{1}{2}x - \frac{3}{4}$
5 $y = A\cos x + B\sin x - \frac{1}{3}\cos 2x$
6 $y = A\cos 3x + B\sin 3x + \frac{4}{37}e^{\frac{1}{2}x}$
7 $y = e^{-2x}(A\cos x + B\sin x) + 2x - 4$
8 $y = e^{x}(A\cos x + B\sin x) + \frac{1}{5}\cos x - \frac{2}{5}\sin x$
9 $y = Ae^{x} + Be^{3x} - x + \frac{2}{3}$
10 $y = A + Be^{-x} - xe^{-x}$
11 $y = A + Be^{3x} - \frac{5}{3}x$
12 $y = Ae^{x} + Be^{-\frac{1}{3}x} + 2 - x$
13 $y = e^{-2x}(A\cos x + B\sin x)$
$+ \frac{1}{65}\sin 2x - \frac{8}{65}\cos 2x$
14 $y = A\cos 4x + B\sin 4x + \frac{3}{2}$
15 $y = e^{-\frac{1}{2}x}\left(A\cos\frac{x}{2} + B\sin\frac{x}{2}\right) + \frac{1}{10}\sin x - \frac{3}{10}\cos x$
16 $y = -\frac{13}{2}e^{x} + \frac{5}{2}e^{3x} + 4$
17 $y = -\frac{1}{2}(\cos x + \sin x - e^{x})$
18 $y = (1 - \frac{1}{2}x)e^{x} - \frac{1}{2}\sin x$
19 $y = \frac{9}{4}e^{x} + \frac{1}{12}e^{5x} - \frac{1}{3}e^{2x}$
20 $y = 2e^{-x}\cos x + 2x - 2$
21 $y = \frac{1}{2}(e^{-2x} + 1)(x + 1)$
22 $y = e^{-x}(\cos 3x + \frac{5}{3}\sin 3x) + 2x - 1$
23 $y = \frac{1}{68}e^{-3x}(4\cos 4x - \sin 4x)$
$+ \frac{1}{17}(4\sin x - \cos x)$
24 $y = \sin 3x + \sin x$
25 $y = e^{6x} - 8e^{-x} + 6x + 7$
26 $y = A\cos x + B\sin x + \frac{1}{2}x\sin x$
27 $k = -\frac{2}{5}$, $y = \frac{1}{25}(e^{12x} - e^{2x}) - \frac{2}{5}xe^{2x}$
28 $y = e^{-x}(2\sin 2x - 4\cos 2x) + 5e^{-x}$
29 $y = Ae^{\frac{x}{4}} + Be^{x} - 4\cos x - \sin x$
30 $y = \frac{1}{2}(\sin 2x - 3\cos 2x) + 2e^{-x}$

Exercise 6E

1 $y^2 = 2x^2 \ln|Cx|$
2 $y = x\ln|Cx|$
3 $Cx^2y^2 = (y - x)^3$
4 $y = \ln|x + y + 1| + \frac{1}{2}x^2 + C$
5 $y = A\cos(\ln|x|) + B\sin(\ln|x|)$
6 $y = Ax^2 + Bx^3$
7 $y = Ax^3 + \frac{1}{2}x^2 + B$
8 $2\ln|3x - y - 3| = y - x + 1 + 2\ln 2$
9 $y = \dfrac{1}{x(C - 2x)^{\frac{1}{2}}}$, $y = \dfrac{1}{x(3 - 2x)^{\frac{1}{2}}}$

Exercise 7A

1 (a) $(3, 0)$ (b) $(4, \frac{\pi}{2})$
(c) $(3.5, \pi)$ (d) $(5, -\frac{\pi}{2})$
(e) $(5, 0.927)$ (f) $(5, 2.498)$
(g) $(\sqrt{2}, -\frac{3\pi}{4})$ (h) $(13, -1.176)$
(i) $(1.5, -2.498)$ (j) $(1, \frac{\pi}{6})$

2 (a) $(3, 0)$ (b) $(0, 2)$
(c) $(0, -5)$ (d) $(-2, 0)$
(e) $\left(\frac{3}{2}, \frac{3\sqrt{3}}{2}\right)$ (f) $(-2, 2\sqrt{3})$
(g) $(2\sqrt{3}, -2)$ (h) $(-2\sqrt{3}, -2)$
(i) $(1.62, -2.52)$ (j) $(-0.83, -1.82)$

6 (a) $(x^2 + y^2)^2 = 4xy$

7 $r + 2\cos\theta = 0$, $r = 2\sin\theta$

8 (a) $r = a\cos\theta$

(b) $r^2 \sin 2\theta = 8a^2$

(c) $r^2 = a^2 \sin 2\theta$

9 (a) $x^2 = ay$

(b) $x^2(x^2 + y^2) = 4a^2y^2$

(c) $(x^2 + y^2 + ax)^2 = a^2(x^2 + y^2)$

Exercise 7B

1 $4.5a^2$ **2** $\frac{1}{4}\pi a^2$ **3** $\frac{1}{4}a^2(4 - \pi)$

4 $2\pi a^2$ **5** $\frac{3\pi}{4}a^2$ **6** $\frac{\pi}{12}a^2$

7 $\dfrac{a^2}{2}$ **8** $\frac{1}{3}a^2$ **9** $\dfrac{a^2}{8}(4 - \pi)$

10 $\frac{11}{2}\pi a^2$

11 (a) $\frac{9\pi}{2}$

(b) (2.37, 1.20), (2.37, −1.20)

13 $\left(e^{\frac{\pi}{4}}, \frac{\pi}{4}\right)$, $\left(e^{\frac{5\pi}{4}}, \frac{5\pi}{4}\right)$

14 (a) $\frac{1}{2}a^2 \ln(2 + \sqrt{3})$

(b) $r\cos\theta = a$

15 (a) $\frac{\pi}{16}a^2$

Review exercise 2

1 $y\sin x = \frac{2}{3}\sin^3 x + \frac{1}{3}\sqrt{2}$

2 $y = Ae^{3x} + Be^{2x} + \frac{1}{2}(\cos x + \sin x)$

3 $\frac{9}{8}\pi a^2$

4 $y = e^{-\frac{1}{2}x}\left(A\cos\dfrac{x\sqrt{3}}{2} + B\sin\dfrac{x\sqrt{3}}{2}\right)$

5 $\ln|y| = x + C + \ln|x - 1|$, $y = \frac{3}{2}e^{\frac{1}{3}}$

6 0.618

7 $y = x\cos x$

8 $y = 4 + Ce^{-x^2}$

9 $\frac{5}{4}\pi$

10 $y = \cos kx + \dfrac{1}{k}\sin kx$

11 $y = \frac{1}{2}x^3 + Cx$

12 $\left(\dfrac{a}{\sqrt{2}}, \dfrac{\pi}{6}\right)$

13 $p = -1$, $q = 0$

$y = 2(2x + 1)e^{-2x} - \cos 2x$

14 (b) 2

(c) $y = 2e^4\left(\dfrac{1}{x} - 1\right)e^{-2x} + \dfrac{2}{x}e^{2x}$

(d) does not remain finite

15 $k = \frac{1}{9}$, $y = e^{2x}(A\cos 3x + B\sin 3x + \frac{1}{9})$

16 2π

17 $y = x\left[\dfrac{\ln x - 2}{\ln x - 1}\right]$

18 $(2a, \pi)$

19 $p = 2$, $q = 0$;

$y = 2\cos x + e^{-4x}(A\cos 3x + B\sin 3x)$

$y = 2\cos x + e^{-4x}(6\cos 3x + 9\sin 3x)$

20 $y\sec^2 x = \tan x - x + C$

21 $\frac{\pi}{6} - \frac{\sqrt{3}}{8}$

22 $y = \frac{1}{2}\cos x(e^{2x} + 3)$

23 $y = Ax^2 + Bx^3 + \frac{1}{2}$

24 $(2\sqrt{2}, \pm\frac{1}{4}\pi)$

25 $y = e^{-x}(A\cos 2x + B\sin 2x)$

26 $y = e^{-x}(A\cos 3x + B\sin 3x)$

(a) $p = 2$, $k = 3$;

$y = e^{-x}(3\cos 3x + 2\sin 3x) + 3e^{3x}$

(b) (i) $y = e^{-x}\sin 3x$

(ii) $e^{-\frac{n\pi}{3} - \frac{1}{3}\arctan 3}\sin(n\pi + \arctan 3)$

27 $\left(\dfrac{2ab}{a + b}, \arccos\dfrac{b - a}{b + a}\right)$; $(\frac{3}{4}\pi - 2)a^2$

28 $y = \sec x[\ln|\sec x + \tan x| + C]$

29 $y = (3x + 2)e^{2x} + x + 1$

30 $y = e^{-2x}(A\cos 3x + B\sin 3x) + \frac{1}{10}e^{-3x}$

31 $y = \dfrac{1}{(1 - 2x + Ce^{-2x})^{\frac{1}{2}}}$

32 $y = \dfrac{2}{x}\sin\dfrac{x}{2} - \cos\dfrac{x}{2}$; $k = 0$

33 $a = -\frac{10}{39}$, $b = -\frac{5}{13}$

$y = A + Be^{-2x} - \frac{10}{39}\cos 3x - \frac{5}{13}\sin 3x$

35 $y = x(Ce^{2x} - 1)$

36 $a = -\frac{1}{2}$, $b = -\frac{1}{4}$;

$y = Ae^{2x} + Be^{-\frac{2}{3}x} - \frac{1}{2}\cos 2x - \frac{1}{4}\sin 2x$

37 (a) $\left(\dfrac{3a}{2}, \dfrac{\pi}{3}\right)$ (b) $\frac{5}{8}\pi a^2$

38 (b) $y = e^{-2x}(A\cos 3x + B\sin 3x) + \frac{7}{5}e^{4x}$

(c) $y = -\frac{7}{5}e^{-2x}(\cos 3x + 2\sin 3x) + \frac{7}{5}e^{4x}$

39 $y^{-2}e^{x^2} = 2x + C$

40 $(1.095a, 1.19)$

41 $y = \dfrac{3 + 2\cos 2x - \cos 4x}{8 \sin x}$

42 (a) $x^2 - y^2 = 1$

(b) $2xy = 1$

43 $\dfrac{d^2y}{dx^2} + 9y = 9x^2 + 20$

44 (a) $r = \dfrac{3}{\cos\theta + 4\sin\theta}$

(b) $r = \dfrac{2\tan\theta}{\cos\theta + \sin\theta}$

45 $y = 3e^{-x} - 2e^{-2x} + 2xe^{-x}$

Examination style paper FP1

1 (b) 0.541

2 $y = x\ln(kx)$

3 (b) $4 - \sqrt{7} < x < 3$ or $5 < x < 4 + \sqrt{7}$

4 (b) $\frac{349}{2520}$

5 (a) $2\pi + \dfrac{3\sqrt{3}}{2}$

(b) $(\frac{1}{2}, 1.82)$, $(\frac{1}{2}, -1.82)$

6 (a) (i) 4, $-2 - 3i$

(b) $p = 2$, $q = 23$

7 $y = A\cos(2\ln x) + B\sin(2\ln x) + \frac{1}{4}\ln x$

List of symbols and notation

The following notation will be used in all Edexcel examinations.

$\in$	is an element of
$\notin$	is not an element of
$\{x_1, x_2, \ldots\}$	the set with elements $x_1, x_2, \ldots$
$\{x : \ldots\}$	the set of all x such that ...
$\mathrm{n}(A)$	the number of elements in set A
$\varnothing$	the empty set
$\mathscr{E}$	the universal set
A	the complement of the set A
$\mathbb{N}$	the set of natural numbers, $\{1, 2, 3, \ldots\}$
$\mathbb{Z}$	the set of integers, $\{0, \pm 1, \pm 2, \pm 3, \ldots\}$
$\mathbb{Z}^+$	the set of positive integers, $\{1, 2, 3, \ldots\}$
$\mathbb{Z}_n$	the set of integers modulo n, $\{0, 1, 2, \ldots, n-1\}$
$\mathbb{Q}$	the set of rational numbers $\left\{\frac{p}{q} : p \in \mathbb{Z}, q \in \mathbb{Z}^+\right\}$
$\mathbb{Q}^+$	the set of positive rational numbers, $\{x \in \mathbb{Q} : x > 0\}$
$\mathbb{Q}_0^+$	the set of positive rational numbers and zero, $\{x \in \mathbb{Q} : x \geqslant 0\}$
$\mathbb{R}$	the set of real numbers
$\mathbb{R}^+$	the set of positive real numbers, $\{x \in \mathbb{R} : x > 0\}$
$\mathbb{R}_0^+$	the set of positive real numbers and zero, $\{x \in \mathbb{R} : x \geqslant 0\}$
$\mathbb{C}$	the set of complex numbers
(x, y)	the ordered pair x, y
$A \times B$	the cartesian product of sets A and B, $A \times B = \{(a, b) : a \in A, b \in B\}$
$\subseteq$	is a subset of
$\subset$	is a proper subset of
$\cup$	union
$\cap$	intersection
$[a, b]$	the closed interval, $\{x \in \mathbb{R} : a \leqslant x \leqslant b\}$
$[a, b)$	the interval $\{x \in \mathbb{R} : a \leqslant x < b\}$
$(a, b]$	the interval $\{x \in \mathbb{R} : a < x \leqslant b\}$
(a, b)	the open interval $\{x \in \mathbb{R} : a < x < b\}$
$y R x$	y is related to x by the relation R
$y \sim x$	y is equivalent to x, in the context of some equivalence relation
$=$	is equal to
$\neq$	is not equal to
$\equiv$	is identical to *or* is congruent to

$\approx$	is approximately equal to
$\cong$	is isomorphic to
$\propto$	is proportional to
$<$	is less than
$\leqslant$, $\ngtr$	is less than or equal to, is not greater than
$>$	is greater than
$\geqslant$, $\nless$	is greater than or equal to, is not less than
∞	infinity
$p \wedge q$	p and q
$p \vee q$	p or q (or both)
$\sim p$	not p
$p \Rightarrow q$	p implies q (if p then q)
$p \Leftarrow q$	p is implied by q (if q then p)
$p \Leftrightarrow q$	p implies and is implied by q (p is equivalent to q)
$\exists$	there exists
$\forall$	for all
$a + b$	a plus b
$a - b$	a minus b
$a \times b$, ab, $a.b$	a multiplied by b
$a \div b$, $\frac{a}{b}$, a/b	a divided by b
$\sum_{i=1}^{n} a_i$	$a_1 + a_2 + \ldots + a_n$
$\prod_{i=1}^{n} a_i$	$a_1 \times a_2 \times \ldots \times a_n$
$\sqrt{a}$	the positive square root of a
$\lvert a \rvert$	the modulus of a
$n!$	n factorial
$\binom{n}{r}$	the binomial coefficient $\frac{n!}{r!(n-r)!}$ for $n \in \mathbb{Z}^+$; $\frac{n(n-1)\ldots(n-r+1)}{r!}$ for $n \in \mathbb{Q}$
$\mathrm{f}(x)$	the value of the function f at x
$\mathrm{f}: A \to B$	f is a function under which each element of set A has an image in set B
$\mathrm{f}: x \mapsto y$	the function f maps the element x to the element y
f^{-1}	the inverse function of the function f
$\mathrm{g} \circ \mathrm{f}$, gf	the composite function of f and g which is defined by $(\mathrm{g} \circ \mathrm{f})(x)$ or $\mathrm{gf}(x) = \mathrm{g}(\mathrm{f}(x))$
$\lim_{x \to a} \mathrm{f}(x)$	the limit of $\mathrm{f}(x)$ as x tends to a
Δx, δx	an increment of x
$\frac{\mathrm{d}y}{\mathrm{d}x}$	the derivative of y with respect to x

$\dfrac{d^n y}{dx^n}$	the nth derivative of y with respect to x
$f'(x), f''(x), \ldots f^{(n)}(x)$	the first, second, ... nth derivatives of $f(x)$ with respect to x
$\int y\,dx$	the indefinite integral of y with respect to x
$\int_a^b y\,dx$	the definite integral of y with respect to x between the limits $x = a$ and $x = b$
$\dfrac{\partial V}{\partial x}$	the partial derivative of V with respect to x
$\dot{x}, \ddot{x}, \ldots$	the first, second, . . . derivatives of x with respect to t
e	base of natural logarithms
e^x, $\exp x$	exponential function of x
$\log_a x$	logarithm to the base a of x
$\ln x$, $\log_e x$	natural logarithm of x
$\lg x$, $\log_{10} x$	logarithm to the base 10 of x
sin, cos, tan cosec, sec, cot	the circular functions
arcsin, arccos, arctan arccosec, arcsec, arccot	the inverse circular functions
sinh, cosh, tanh cosech, sech, coth	the hyperbolic functions
arsinh, arcosh, artanh, arcosech, arsech, arcoth	the inverse hyperbolic functions
i	square root of -1
z	a complex number, $z = x + iy$
Re z	the real part of z, Re $z = x$
Im z	the imaginary part of z, Im $z = y$
$\lvert z \rvert$	the modulus of z, $\lvert z \rvert = \sqrt{(x^2 + y^2)}$
arg z	the argument of z, arg $z = \arctan \dfrac{y}{x}$
z^*	the complex conjugate of z, $x - iy$
$\mathbf{M}$	a matrix $\mathbf{M}$
$\mathbf{M}^{-1}$	the inverse of the matrix $\mathbf{M}$
$\mathbf{M}^{T}$	the transpose of the matrix $\mathbf{M}$
det $\mathbf{M}$, $\lvert \mathbf{M} \rvert$	the determinant of the square matrix $\mathbf{M}$
$\mathbf{a}$	the vector $\mathbf{a}$
$\overrightarrow{AB}$	the vector represented in magnitude and direction by the directed line segment AB
$\hat{\mathbf{a}}$	a unit vector in the direction of $\mathbf{a}$
$\mathbf{i}, \mathbf{j}, \mathbf{k}$	unit vectors in the directions of the cartesian coordinate axes
$\lvert \mathbf{a} \rvert$, a	the magnitude of $\mathbf{a}$
$\lvert \overrightarrow{AB} \rvert$, AB	the magnitude of $\overrightarrow{AB}$

$\mathbf{a} . \mathbf{b}$ the scalar product of **a** and **b**
$\mathbf{a} \times \mathbf{b}$ the vector product of **a** and **b**

A, B, C, etc events
$A \cup B$ union of the events A and B
$A \cap B$ intersection of the events A and B
$\mathrm{P}(A)$ probability of the event A
A complement of the event A
$\mathrm{P}(A|B)$ probability of the event A conditional on the event B
X, Y, R, etc. random variables
x, y, r, etc. values of the random variables X, Y, R, etc
$x_1, x_2 \ldots$ observations
$f_1, f_2, \ldots$ frequencies with which the observations $x_1, x_2, \ldots$ occur
$\mathrm{p}(x)$ probability function $\mathrm{P}(X = x)$ of the discrete random variable X
$p_1, p_2, \ldots$ probabilities of the values $x_1, x_2, \ldots$ of the discrete random variable X
$\mathrm{f}(x), \mathrm{g}(x), \ldots$ the value of the probability density function of a continuous random variable X
$\mathrm{F}(x), \mathrm{G}(x), \ldots$ the value of the (cumulative) distribution function $\mathrm{P}(X \leqslant x)$ of a continuous random variable X
$\mathrm{E}(X)$ expectation of the random variable X
$\mathrm{E}[\mathrm{g}(X)]$ expectation of $\mathrm{g}(X)$
$\mathrm{Var}(X)$ variance of the random variable X
$\mathrm{G}(t)$ probability generating function for a random variable which takes the values 0, 1, 2, ...
$\mathrm{B}(n, p)$ binomial distribution with parameters n and p
$\mathrm{N}(\mu, \sigma^2)$ normal distribution with mean μ and variance σ^2
μ population mean
σ^2 population variance
σ population standard deviation
$\bar{x}$, m sample mean
s^2, $\hat{\sigma}^2$ unbiased estimate of population variance from a sample,

$$s^2 = \frac{1}{n-1}\sum(x_i - \bar{x})^2$$

ϕ probability density function of the standardised normal variable with distribution $\mathrm{N}(0, 1)$
Φ corresponding cumulative distribution function
ρ product-moment correlation coefficient for a population
r product-moment correlation coefficient for a sample
$\mathrm{Cov}\,(X, Y)$ covariance of X and Y

Index